It Doesn't Matter Where You Sit

It Doesn't Matter Where You Sit

Fred McClement

Holt, Rinehart and Winston

New York Chicago San Francisco

An earlier edition of this book
was published in Canada by
McClelland and Stewart, Limited.

Library of Congress Catalog Card Number: 69–16187

Published, June, 1969
Second Printing, September, 1969

Designer: Robert Reed

SBN: 03–076510–2

Printed in the United States of America

Without the assistance of the following
agencies, this book would not have been possible:
the Safety Board of the U.S. Department of Transportation,
the Federal Aviation Administration, the Port of
New York Authority, the Air Line Pilots Association,
the British Air Ministry, the Department of Transport
of the Dominion of Canada, the Canadian Air Line
Pilots Association, the Aviation/Space Writers
Association, the British Aircraft Corporation,
Sud Aviation of France, Boeing, Lockheed, Douglas,
United Air Lines, American Airlines, Trans World
Airlines, Air Canada, Canadian Pacific Air Lines,
Rolls-Royce, General Electric, Pratt and Whitney,
the U.S. Air Force, the Royal Canadian Air Force,
the U.S. Navy, a host of commercial air lines
pilots, and a number of newspaper reporters concerned
with the events in this book.

Contents

Foreword

"I love you, Mary . . . love you so much . . . this is goodbye."

These were the last words transmitted over his radio by a pilot in the final anguished moments of his life. As he spoke, his giant four engine airliner fell lazily to the ground, having been struck a moment before by another aircraft.

Though the tape of this terrible drama is now a historical record, the cause of the midair collision that determined the pilot's death is still with us, in every part of the sky highway. Despite all efforts to bring air traffic under some kind of order, chaotic airspace conditions have multiplied over the past several years. In fact, the worst air collisions in the country have occurred in Federally controlled airspace under the reputedly watchful eyes of radar.

Midair collisions and near collisions are increasing at an alarming rate, so much so, that the Federal Aviation Administration has stopped revealing the incidence of near misses. And no wonder. After granting immunity from reprisal to any pilot reporting a near miss, the FAA was besieged by reports of 554 near collisions in the first ten weeks of 1968, which would seem to indicate an annual average of between 2,000 and 2,500 such incidents. These frightful statistics dramatize the fact that air traffic in the United States is apparently in chaos, and that it is only Dame Fortune who keeps the death statistics down . . . for the moment.

Who is to blame? Government regulating agencies, airport operators, the airlines themselves, and metropolitan planning boards. However, being able to establish the blame for the United States air safety record is certainly no comfort to the next of kin of the seventy-five persons who died near Hendersonville, North Carolina, on July 19, 1967, when a small Cessna aircraft "deviated into the airspace allocated to a Piedmont 727 jet."

Like most United States airports, the Asheville (North Carolina) Airport, which was handling the control of aircraft near Hendersonville that day, had no radar to keep track of the traffic in the vicinity. Like many other airports in the country it is rated by the Air Line Pilots Association as "below standard" in fire and rescue equipment, while its main jet runway is almost three thousand feet below the minimum recommended length.

Nor is trying to place the blame on the jet pilots any balm to the next of kin of those killed in a midair collision just outside Dayton Airport on March 9, 1967. In that crash, a DC-9 of Trans World Airlines was making an approach in a recognized air corridor under Federal control when it collided with a small plane. Twenty-six lives were destroyed. This incident led the Safety Board of the U.S. Department of Transportation to comment: "The lack of positive control over aircraft operations conducted in terminal areas under the present day air traffic control system is not satisfactory." That comment was made in August 1968 not, as you might suppose, in August 1933.

Of the 554 reports of near disasters in the first ten weeks of 1968, some 160 involved commercial airliners; 339 of these near misses happened in the immediate vicinity of airports, while the remainder occurred elsewhere along the sky highways. From January 1 to October 1, 1968, there were thirty-four midair collisions, with heavy loss of life.

What do these incidents indicate?

They show dramatically that United States airspace is dangerous; air traffic control is *out* of control; there are too many aircraft, big and small; there is insufficient airspace around the air terminals of great cities. There are congested air terminals, a scarcity of radar surveillance, poor aircraft instrumentation, bad maintenance of jet airliners, constant infringement of safety rules, outmoded airports and safety equipment, slipshod weather forecasting, and, often, questionable commercial flying techniques. Combinations of these factors are revealed with nagging regularity *every time* there is an air disaster.

Aviation is in such a hurry it is skidding into public unacceptability. There is no better way to describe it. The problems of the first decade of the jets have not been solved sufficiently at this date to permit the introduction of jumbojets that will carry between three hundred and one thousand passengers; nor is the United States ready for supersonic airliners that will whiz by in a storm of sonic explosions at two or three times the speed of sound.

However, the jumbojets are already here and the supersonic British-French Concorde is being tested. The Russian Tupolev has already flown. Yet, at this very moment, the Department of Transportation's Safety Board knows, and has published the fact, that there are jet airliners in the air right now that are not strong enough to withstand the gusts of wind found in ordinary, everyday thunderstorms. To try to overcome this situation, the Safety Board in mid-1968 urged the FAA to make it mandatory for United States air carriers to circumnavigate severe storm areas, or, if necessary, to wait on the ground until the storms passed by.

The Safety Board letter to William F. McKee, Administrator of the Federal Aviation Administration, read in part:

> The Safety Board has become increasingly concerned with aircraft accidents which have occurred during periods of severe weather conditions and we are particularly concerned with the catastrophic air carrier-type accident.
>
> Accordingly, in the interest of aircraft accident prevention, the Safety Board recommends that consideration be given to the following proposed action:
>
> 1. Amend Parts 11 and 135 of the Federal Aviation Regulations to prohibit penetration of thunderstorm activity occurring in an area for which tornadoes and/or severe thunderstorms have been forecast, and
>
> 2. Require that, as an interim measure until standards have been developed, these storms be avoided by at least 20 nautical miles.
>
> The Safety Board feels very strongly that over the years since airborne weather radar has been in use on a routine basis there has been a tendency for that equipment to be utilized in a manner that, to our knowledge, was not intended. The equipment was to be used as "Weather Avoidance" radar, rather than for deliberate penetration. Recently we conducted a survey of a representative group of air carriers to obtain information on the general policies and methods of operation during periods of severe weather. All

of these carriers do have a policy of thunderstorm avoidance rather than penetration. Nevertheless, penetrations occur and accidents take place.

Our concern is so great that in addition to recommending regulatory action we plan to write a letter to all air carriers stressing the need for all personnel to adhere to published policies and procedures with emphasis placed on the use of airborne weather radar for storm *avoidance* purposes.

The FAA turned down the request, leaving the decisions as to navigation of flight around storms and as to the timing of takeoffs, to the airlines and more particularly to individual aircraft captains and airline dispatchers. Up to now these men have shown a preference for following company schedules rather than heeding the warnings of severe weather forecasts.

The Safety Board tries, but how can it possibly win the battle of air safety? Study, for instance, the aborted takeoff crash of a TWA Boeing 707 at the Greater Cincinnati Airport on November 6, 1967. Even the Safety Board didn't agree on who was to blame for this mishap. The aircraft was rolling down the runway for its takeoff when the copilot, on the right-hand side of the aircraft, thought that he had struck a Delta DC-9 airliner which was mired just off the runway. He was unaware that a "shock blast" from the DC-9 jet engines had slammed against his 707 and had actually caused a momentary disruption in the power output of one of his engines.

The copilot pulled off the power of the hurtling 707, slammed on the brakes, and threw the throttles into reverse. With a thunderous roar, the jet went over the end of the runway and broke up at the airport perimeter. Eleven of the thirty-six passengers were injured. One died four days later.

The blame was eventually placed on the crew for their inability to abort successfully. But let's look at the improbable odds of successfully aborting a jet flight at a present-day airport.

The recommended jet runway length at Cincinnati is 10,600 feet. The runway being used on that takeoff by the TWA was only 7,800 feet in total length, and by the time it reached the DC-9, the jet had already covered a considerable part of that distance. It has been established by both Boeing and by the Safety Board that a 707 hurtling along the ground at 143 knots requires a stop-distance of 7,850 feet, or else, if something goes wrong, it will be impossible to avoid a smashup. The Cincinnati runway is too short. Almost all United States jet airports today have inadequate runways.

So, fasten your seat belts. Although it doesn't matter where you sit, breakup of the fuselage, such as occurred in this crash, may mercifully open enough metal to permit you to escape. If the fuselage doesn't peel open, then the next step is to fight your way through screaming passengers—men, women, children, oldsters, and babies in arms—to try to find one of the pitifully few exits provided to save your life.

This book is dedicated to safety. Your safety. There may still be time.

March, 1969 Fred McClement

It Doesn't Matter Where You Sit

1

"They're Going Down in Flames"

Along the network of modern highways that link Baltimore, Maryland, to the cities of Wilmington, Delaware, and Philadelphia, Pennsylvania, the driving conditions on that night of December 8, 1963, were hazardous, to say the least.

Driving winds and blinding sheets of rain mixed with fog and illuminated by brilliant displays of electrical pyrotechnics swept across the rolling countryside with a fury that is most often associated with the thunderstorms of a hot summer. But the weather that December had been unusually mild and, because of that, thousands of motorists had been lured into the pulsing arteries of the mideastern seaboard. On that Sunday night they were homeward bound when they were trapped by the great storm.

The bad weather was being generated by a squall along the leading edge of a vast cold front which had originated over Canada. Accompanied by a solid line of severe thunderstorms, it was moving across the country toward the Atlantic. The U.S. Weather Bureau predicted earlier in the day that the storm would reach the areas of Baltimore and Philadelphia at eight o'clock and it arrived right on time. In the little community of Elkton, Maryland, nestled among the concrete highway complexes, most residents were content to remain indoors and watch Ed Sullivan, though a few teenagers headed for the town's indoor skating rink undaunted by the lightning and rain. But for those thousands of motorists, the going was rough and, during the period when visibility deteriorated to a few hundred feet, they pulled to the shoulders of the roadways to await an improvement in the weather.

It seems incredible that in a storm of this scope and fury ninety-nine persons would be witnesses to one of the country's worst air disasters. This perplexing situation can be attributed to the fact that the storm was one of the worst in living memory. Those who peered from their doorways and windows into the brilliant panorama in the sky and those who were forced to pull to the roadsides were fated to be the audience at an appalling drama.

From time to time the sounds of great airliners would reach their ears, and though they would search the lightning-filled clouds they were unable to see the scores of aircraft heading for landings at Philadelphia, Wilmington, and beyond. If they thought the storm was bad on the ground they should have been upstairs with the hapless airline passengers. Aerodynamics being what they are in the twentieth century, the giant aircraft overhead were not blessed with the technical ability to pull over, park, and await the passing of the cold front. Furthermore, these planes were flying the densest air corridor in the world, made more busy by the extent of the storm. Because of this they were denied the chance to change their flight paths to find smoother flying conditions.

Not a few veteran air travelers wished they had taken the train or delayed their flights, particularly those passengers who were stacked at various turbulent altitudes awaiting the chance to land at Philadelphia's International Airport.

They were flying round and round in tight circles, with the muffled roar of piston engines in their ears or the high-pitched whine of jet power plants contributing to their anxiety. The aircraft were enveloped by cascades of brilliant lightning and thunderclaps pounded in the passengers' ears. Plane travel was certainly not enjoyable that night. Yet, here the travelers were, strapped tightly in their seats because of the bone-jarring, roller-coaster convection currents. Only an occasional glimpse of tiny lights far below bolstered their courage. They must have wondered what they were doing flying in the middle of all this nasty weather when they had so often been lured by airline advertising that promised them "above the weather" comfort along a radar-controlled flight path.

The eighty-one passengers and crew members of a Pan American 707 Jet Clipper, now taking off from Baltimore for the short hop to Philadelphia, wouldn't have long to wonder . . . just thirty-two minutes and fifteen seconds.

As soon as they were airborne, the Clipper passengers could clearly see the storm ahead. There was no mystery about its presence. Three hours earlier, the Weather Bureaus along the coast had issued storm warnings to all airlines operating in the eastern part of the country, advising that moderate to severe turbulence would be encountered in all levels of the atmosphere over the states of West Virginia, Maryland, Delaware, Virginia, and the northeast section of North Carolina. The District of Columbia was included in the advisory and, for the many planes flying between Florida and New York, the Atlantic coastal waters as well.

No aircraft flying over the area that night, radar guided or otherwise, could possibly escape a storm. At lower altitudes where business and pleasure aircraft were flying, moderate to severe turbulence was predicted; this warning was bleating out at regular intervals over airway frequencies. Severe turbulence was forecast between 12,000 and 25,000 feet in the area where commercial piston and turboprop airliners travel the skies. Above 25,000 feet, in the altitudes of the jets, severe thunderstorms with turbulence were predicted up to 40,000 feet; and in the clear areas between and around these great storm systems clear-air turbulence was forecast. All this bad weather, of course, was typical of a rapidly moving cold front colliding with warm and unstable moist air flowing northward.

This particular storm, to add torment to trouble, had a high altitude jet stream associated with it. As a result the entire air mass was more turbulent than had at first been realized. For good measure, tornadoes were predicted. It was a perfect night to stay at home.

Yet at precisely eight o'clock over the Maryland-Delaware countryside, eight planes were circling through the clouds awaiting their chance to descend under precise radar control to the runways of the Philadelphia Airport. These aircraft were stacked in four "holding fixes" some twenty miles to the south and to the west of Philadelphia, holding areas known to pilots and controllers by the towns directly underneath them, New Castle, West Chester, Frazer, and Woodstown.

In the Philly Air Route Traffic Control at that moment was Paul Alexy, the Federal Aviation Agency's Approach Controller, responsible for the shepherding of these eight aircraft and others soon to be arriving in his zone of command.

Such stormy nights were nerve-racking to FAA controllers like Alexy. Watching the eerie green glow of the radar pattern on the approach radar screen as it singled out the little blips of aircraft between the clutter of precipitation was bad enough, without the sound of many voices in the system and the jarring slams of lightning discharges in the earphones. Trying to keep the aircraft in the four holding sections away from each other without the knowledge of their altitudes—except what was called out by the pilots themselves—was a tough job for even the most experienced controllers in the business. It was Alexy's job to bring each of the eight aircraft from the holding fixes to the safety of the gleaming runways, and he could accomplish the task only by singling out one plane at a time with his radar and guiding it with his voice down through the clouds and rain to an area south of the airport. Then he would turn the aircraft into a heading to the runway, where the control tower would take over the final descent and landing. Alexy was perspiring and no wonder, with all eight planes anxious to land and a big 707 just reported by the Air Route Traffic Control to be coming into the Philadelphia holding complex. That would be the Pan Am from Baltimore, known as Clipper 214.

The number of aircraft stacked up at that moment was

not unusually high for a metropolitan airport on a stormy, fog-filled night. Nor was it unusual that the various planes spanned two decades of air progress. There was an Allegheny Airlines Martin 202; an Aero Commander twin-engine private plane; two Allegheny Convair 440s; two four-engine DC-7s, operated by United Air Lines and by Eastern Airlines; a United Air Lines Viscount; a National Airlines DC-8 jet; and the Pan Am 707.

The wind was thirty knots from the west, which meant that landings and takeoffs from the airport would be made at a heading of 270 degrees along Runway 27.

Arriving planes would descend to 3,000 feet and circle in such a manner as to approach the airport from the east or southeast, while departing airliners would take off into the west wind, turning sharply northwest and then northward to keep well away from arriving aircraft.

Heavy showers cluttered Alexy's set and rivers of lightning bathed the night with brilliance, drowning out communication and blinding the pilots at the controls of their aircraft. But through it all, the radar signal pulsed in milliseconds and tried to pick out each plane and reduce it to a tiny blip on a circular screen for the controller to study.

Alexy kept up a running chatter with each of his flock on the constant changes in the weather. Without radar, landing the heavy jets under such squally conditions would have been almost impossible. Because of their great speed, the job of tracing their circling patterns was continuous and tiring. But the fact that jets are susceptible to stalling in turbulence if their speed is reduced below two hundred miles an hour made it impossible to slow them down. As it was no secret that altimeters had failed on too many occasions, Alexy queried the pilots continually to make sure of their aircraft's height while he handled lateral separations.

Alexy was very much concerned with the movement of the squall line and was not anxious to guide any of the aircraft into the final approach until the winds and heavy rains had moved out of range. Pilots and passengers would be irritated by the lengthy landing delays, and complaints from the housing developments under the holding areas would start any minute now, but Alexy was being careful. He already had reports of severe turbulence in one of the fixes and he would take no chances on the possibility of a goofed-up landing.

No matter what the weather, Alexy's job was anything but humdrum. In fact, frequent rest periods were required for all controllers because of the tremendous pressure of handling high-speed aircraft in high-density complexes such as Philadelphia. During bad weather, like that night's, controllers were always faced with the specter that one slip could cause a midair collision or slam an airliner into a hill just short of the runway. So Alexy's eyes were glued constantly to the radarscope as the little blips of airplanes identified by tiny numbered markers called "shrimp boats" moved in unison across the scope.

One of his shrimp boats identified National Airline's Flight 16, a Douglas DC-8 jet circling at 6,000 feet in the New Castle holding area directly over the invisible boundary that separates Delaware from Maryland.

At the controls of the mighty jet were two veterans of National Airlines, Captain Malcolm M. Campbell, the plane's commander, and Captain Gerald Sutliff, his copilot. Campbell with twenty years of service and 1,150 hours in the big jets sat in his traditional place on the left-hand side of the flight deck. He was concentrating on keeping his airliner under the exact power settings needed to hold it as stable as possible in the bumpy storm clouds. Sutliff, who had been flying commercial airliners since 1951, sat on the right side and was assisting in handling the controls as well as keeping up a steady conversation with Alexy over the progress of the storm and the location of other aircraft hidden in the clouds.

Campbell and Sutliff were concerned about this storm. Their airborne radar showed them several thunderstorm cells in the vicinity, but it could not pick out the turbulence created by sporadic wind gusting and they were at a low altitude where such gusting is usually found. Lightning filled the sky around them, and reports of turbulence from other aircraft made the situation on the National flight deck tense. Neither captain had any intention of keeping to their flight schedule by tangling with that churning cold front in the vicinity of the airport. The minutes droned on as they circled through the night.

After a short interval, Alexy returned his attention to this flight and gave National permission to leave the area and head for the airport by a route that would take it east of Wilmington for a gradual left-hand turn to the runway—every step under precise radar vectoring. It was 8:00 P.M. and over the crackling radio the permission sounded like this: "Let's see, National 16, we haven't forgotten you . . . depart the New Castle VOR, heading three six zero . . . maintain 6,000 . . . this will be a radar vector to the ILS final approach course . . . wind 270 degrees, at 25 knots, indicating Runway 27."

Alexy was telling Sutliff to fly the DC-8 out of the holding pattern on a course, north by northwest, for an Instrument Landing System Approach to the active runway. Sutliff answered him: "Okay . . . what is the ceiling and visibility there?"

Alexy replied: "Here's the Philly weather . . . scattered clouds at 600, 900 broken . . . and, oh . . . 2,000 overcast . . . Two and a half miles, rain and showers . . . altimeter, they give us now, two nine four eight."

This forecast indicated that the base of the storm cloud was now high enough and the ground visibility lengthy enough to try a landing and Sutliff answered: "Okay and thanks."

Another voice cut in. This was the pilot of an Allegheny Airlines Convair 440 which was rapidly approaching the New Castle holding area just a little above the circling National jet. The flight, known to the Philly controller as Flight 908, asked for an immediate clearance to land. Being a two-engine piston aircraft, it was not subject to the same ceiling minimums as the big jets and it was also able to throttle down to approximately 125 miles an hour for a slow approach through the weather to the airport. Alexy told Allegheny 908 to start circling at 8,000 feet until he was ready for it. The conversation between them over the weather made the National pilots undecided whether to hold their position or go on in. From long years of experience they didn't "feel right" about the conditions of the air between them and the airport and, because they were free of turbulence at the moment, they were not too anxious to leave the relative security of the holding area. They were unaware that the narrow section in which they were circling was soon to be the most dangerous of all.

Captain Sutliff radioed: "What indications have you for us whether that weather will improve at your airport shortly?" Alexy replied that he would ask the Weather Bureau.

At the very second that Flight 16 was asking for an updated report, the United States Weather Bureau's teletypes were being fed a severe storm warning for the precise areas in which these airliners were circling: OVER EASTERN MARYLAND, EASTERN VIRGINIA, DELAWARE, EAST NORTH CAROLINA AND COASTAL WATERS . . . FREQUENT MODERATE-TO-SEVERE TURBULENCE BELOW 15,000 FEET WITH LOCALLY EXTREME TURBULENCE BELOW 7,000 FEET.

The National Flight was not advised of this bulletin, but it is doubtful if anything could have been done at the moment to help its situation. Flying out of an encircling storm is just as dangerous as flying into one, so the circumstances indicated that National should remain at 6,000 feet, keep a tight radar watch, and listen to the controller and to the conversations of other pilots.

This latest Weather Bureau announcement was serious enough to be considered a "flash advisory," which would indicate extreme urgency. But for some reason the Bureau did not make this a flash advisory, although the weather had all the ingredients for one—meaning extreme danger to all aircraft at every height in the atmosphere. This bulletin not only predicted moderate turbulence, which was bad enough, but it predicted severe turbulence, which is the weatherman's description of convection currents and wind violence that can pitch an airliner out of control.

Unaware of the severe turbulence warning because it was never mentioned in the radio transmission, Alexy told Allegheny Flight 908 that it would be given clearance to leave the holding pattern at 8:35 P.M.—in about twenty minutes' time. Next he called Eastern Airlines, Flight 4, telling it to hold at 3,000 feet over New Castle and that, immediately after the Allegheny started its approach, Eastern would be next in line.

Then Alexy turned to Captain Sutliff. "Ah, National 16 . . . and all aircraft . . . we're making circling approaches to Runway 27 now . . . winds 270 degrees . . . at 30 knots . . . and I'm getting the latest forecast from the weather . . . okay . . . Weather Bureau advises that the squall line is estimated to pass in ten or fifteen minutes with considerable decrease in wind at that time."

Sutliff requested a repeat of this message and then asked if that squall line was right at the airport at this moment. Alexy replied: "It appears to be . . . in fact, according to the radar, it appears to have just passed, going from west to east . . . however, the Weather Bureau says it will be passing in about ten or fifteen minutes . . . but . . . according to the precipitation on my radar, the heaviest precipitation has already passed, and, National 16, I am awaiting your report on leaving New Castle."

Campbell and Sutliff getting conflicting reports on the whereabouts of the squall line were not anxious to strike severe wind gusts at so low an altitude as 6,000 feet, particularly while negotiating a final approach with their eyes glued to the instrument-landing dials. So Sutliff radioed the airport: "Ah, we're just making one more circle before starting up there . . . to give that squall line a chance to ease out . . . and we'll be leaving New Castle in two or three minutes."

Alexy now turned his attention to the Aero Commander, holding at Woodstown eighteen miles west of the airport at an altitude of 6,000 feet. He gave the pilot the latest weather at the field and said he would be cleared to land at 8:40 P.M. This message was acknowledged. Then the Aero Commander pilot had this to say: "I've slowed down quite a bit . . . uh, we're in some pretty rough turbulence over here, so I've slowed down as much as I can."

"Okay, that's fine," replied Alexy. "There seems to be turbulence in all quadrants right now."

National's pilots heard these significant words while making their final circle, and Sutliff again radioed: "Philly Approach . . . we're not too anxious to land in that squall line right now . . . you say it will be out of there in another ten minutes?"

"Right, National 16 . . . that's fine . . . maintain 6,000 and hold New Castle as directed."

Campbell and Sutliff agreed. Their decision would provide them with a frightening experience in just a few minutes, as the drama in the skies continued.

Alexy called Eastern Airlines Flight 4: "Okay . . . you want to make an approach?"

"Yes, sir, anytime," Eastern politely replied through the static.

This flight was then given permission to leave the 3,000-feet altitude in the New Castle holding position for a radar-controlled approach to the airport. At the same moment, Allegheny's Flight 929 was moved from 8,000 feet to 7,000 feet. Then the controller asked National 16 if it would like to take a try now, and Sutliff replied: "Ah, we'll stay here at six and wait until that line gets by . . . keep us advised please, sir."

Eastern was cleared downward to 1,800 feet on a course north by northeast toward the airport.

Allegheny Flight 908 was to follow, the planes keeping ten miles apart in the landing pattern. The Aero Commander was now ordered directly south and slightly east of the airport to circle in left-hand turns until cleared for final approach. The Allegheny 908 descended to 4,000 feet. Eastern Airlines was seven miles from the first in a series of three markers to the runway and was turned over to the field control tower, relieving Alexy of one aircraft to worry about.

National called in again and asked if the squall line had passed, and the reply came quickly through the storm: "We have some light precipitation, which is east of New Castle, and for New Castle itself we are checking . . . no precipitation. The heaviest line, which we believe to be the squall line, is now about seven miles east of Philadelphia. It's a line running north and south, moving eastward."

The controller again turned his attention to Allegheny Flight 929, which had descended to 7,000 feet. He gave it clearance to approach the airport on a north-northwest heading with a series of turns to the airport. At this moment United Air Lines Viscount Flight 328 was handed over to Alexy and this flight informed him that it was at 5,000 feet and approaching the Philadelphia holding areas.

This was no sooner acknowledged than the Pan American Clipper also arrived in the Philadelphia Approach Control area. It was apparent that the severe weather bulletin had little effect on the continuing movement of aircraft in the dangerous zone.

The Pan Am jet, incidentally, was the first 707 that ever flew commercially. Aboard it were seventy-three passengers, many of whom were heading home after vacations in the West Indies. The flight had left San Juan's International Airport in Puerto Rico at 4:10 P.M. and had made one stop en route at Baltimore, where seventy-one of the original 144 passengers had deplaned.

Identified as Flight 214, it had then taken off from Baltimore's Friendship Airport at 8:25 P.M. for the short hop to Philadelphia and the termination of the trip. At the controls were veteran pilots, Captain George Knuth of Huntington Station, Long Island, and First Officer John R. Dale of Port Washington, Long Island. On the flight deck with them were engineer John Kantlehner of Brentwood, Long Island, and Second Officer Paul Orringer of New Rochelle, New York. Taking care of the passengers were two pursers, Mario Montilla and Joseph Morett, and two stewardesses, Tommie Louise Simms and Virginia Ann Heinzinger, all from the New York region.

The passengers were not being served at this moment, as the seat-belts lights would be on and the 707 would be swooshing through the turbulence and lightning-filled skies into the New Castle holding area. Its expected landing time of 8:45

P.M. was delayed by the storm and by the other aircraft stacked up in the skies. As a result, Flight 214 began to circle with the others at an altitude of 5,000 feet, just 1,000 feet below National 16.

At 8:25 P.M., about the time the Pan Am Clipper was roaring off the Baltimore runway, the Weather Bureau issued yet another bulletin covering the mideastern zone, and it warned: OVER WEST VIRGINIA, MARYLAND, DELAWARE, VIRGINIA, EAST NORTH CAROLINA AND COASTAL WATERS . . . MODERATE TO GREATER CLEAR AIR TURBULENCE FROM 20,000 FEET TO 40,000 FEET . . . ADDITIONAL ADVISORY FOR LIGHT AIRCRAFT, ICING BELOW 10,000 FEET.

The Pan Am Clipper began its circling over New Castle seventeen minutes after this forecast, making right-hand turns on a heading that was westward of New Castle, turns that would take it directly over Elkton, Maryland.

The Clipper's First Officer John Dale called Alexy that his flight was holding at the assigned position.

"All right, Clipper 214 . . . hold west of New Castle on the two seven zero radial, right turns . . . and, ah, here's the Philly weather now . . . 700 scattered . . . 800 broken . . . 1,000 overcast . . . six miles with rain showers . . . altimeter two nine four five . . . the surface wind is 280 degrees at 25 knots with gusts to 30. . . . I've got five aircraft elected to hold until this . . . oh, extreme winds have passed . . . ah, do you wish to be cleared for an approach or would you like to hold until the squall line passes Philly . . . over."

The answer came quick and to the point: "Clipper 214 . . . we'll hold."

At this instant, Allegheny Flight 908, at 4,000 feet, radioed that it would like to come on in. Probably because of the heavy static and so many voices jamming the air, the controller thought it was Allegheny 929 calling. He wondered for a second why this flight would be wanting to make its approach when it was supposed to be over New Castle and stacked up with the others. The confusion was cleared a moment later, and Flight 929 was given permission to descend to 7,000 feet and depart the holding area, while the other Allegheny Flight—908—was ordered to hold at its assigned position of 4,000 feet. Slight mix-ups like this often occur in conditions where heavy static drowns out the completion of messages. But they are usually quickly cleared up. In this severe storm, however, anything was likely to happen.

Meanwhile the two jets keep circling—the National at 6,000, and the Pan Am Clipper at 5,000. Permission from the controller was given the Clipper to extend its circle into a two-minute pattern instead of a one-minute oval. This would give it more range to circle and provide better controllability if severe turbulence should be encountered. Pilots know how dangerous it is to bank too sharply in turbulence. They can find themselves in a downward power dive which could be impossible to control.

Allegheny 929 was cleared to 1,800 feet and was ten or twelve miles from the outer marker, which was five miles from the end of the runway. It was told that there was a band of precipitation six miles long through which it would have to fly. The controller ordered Allegheny's Flight 908 out of the holding pattern on a course that would take it over Wilmington, visible only by its glow in the heavy rain clouds.

Then the Pan Am Clipper radioed Alexy: "Clipper 214, we're ready to start an approach."

"Okay, Clipper 214 . . . hold as instructed and I'll pull you away as soon as I can."

"Ah, no hurry, just wanted you to know that we'll accept a clearance."

"All right, that's fine."

National 16 was still 1,000 feet above the Clipper. Its pilots were silent as they listened to the conversations, still holding their giant DC-8 at the recommended speeds for turbulence penetration and content to await better weather.

The recommended speed for the Pan Am 707 for stack-up in holding fixes was 230 knots over 14,000 feet; 210 knots from 6,000 to 14,000; and 200 knots under 6,000 feet. The

Boeing people had only recently recommended that the 707 series increase their turbulence speed from the previous range of from 220 knots and 250 knots to an air speed of 270 knots. This was to prevent the possibility of stalling caused by wind gusts and the loss of control which could follow.

So, while the 707 Clipper circled according to the book, the controller turned his attention to Allegheny 908 and the Aero Commander, both in their final approaches to the airport. When these conversations were completed, National 16 called in: "Uh, what are you showing now on the weather there, please?"

Alexy replied: "Ah, here's the latest weather and, uh, for all aircraft holding . . . Philadelphia weather, 700 scattered, ceiling 800 broken, 1,000 overcast, six miles with rain shower, surface wind is 330 degrees at fifteen, Runway 27."

This weather picture looked a little brighter, with the ceiling at 800 feet, the winds down to fifteen knots, and the visibility six miles in the rain. A few minutes later, the controller informed National 16 that he was bringing down two more aircraft, one from the Frazer holding area and the other from West Chester, and that National would be next in line.

"Things are looking better," he said cheerfully.

But National's pilots did not get the message clearly because of the increased static from the fountains of lightning around them, and the message was repeated.

"Okay," said Sutliff.

"And how's the turbulence now in your area, National 16?" asked the controller.

Before National could answer, a voice came out of the air, chilling the hearts of everyone who heard it. It came from First Officer John Dale in the Pan Am Clipper, circling at 5,000 feet: "MAYDAY . . . MAYDAY . . . MAYDAY . . . CLIPPER TWO ONE FOUR OUT OF CONTROL . . . HERE WE GO."

Alexy gripped his microphone and yelled: "Clipper 214, did you call Philadelphia?"

National's Sutliff replied instead, as he stared with unbelieving eyes at the sight just ahead and a thousand feet below him: "CLIPPER TWO FOURTEEN IS GOING DOWN IN FLAMES."

The controller immediately cleared all air traffic beneath the plunging jet. The two National pilots were stunned by the sight of the ball of fire diving earthward. Then their own airliner was struck by a brilliant arc of lightning. They were shaken by the force and their immediate reaction was to get out of the area as quickly as possible and abandon the thought of getting through the storm into the Philadelphia airport.

"Ah, Philadelphia, this is National 16 . . . we'd like to get up to the New York area . . . we'll continue to Newark or Idlewild."

Sutliff was so shaken by his experience (it was the first time the two pilots had ever heard a "Mayday" or distress call) that he at first said Woodstown instead of Philadelphia and then asked about the weather at Philly when he meant Newark and Idlewild.

The controller called back through the static: "All right National 16 . . . I have traffic holding at Woodstown at 6,000 and 7,000 feet and, ah, we'll stand by and see if we can get a higher altitude at New Castle."

"We don't want to stay here," snapped Sutliff. Who could blame him? He had just witnessed a giant airliner going down in flames while his own jet was struck by a vicious bolt of lightning. He might even be on fire or have sustained damage, although the instruments did not report anything unusual.

"Roger . . . understand National 16 . . . ah, turn right . . . take a one eight zero heading out of the New Castle area . . . ah, this could put you in a possible smoother area."

"It's smooth enough here . . . we're just getting lightning."

The conversation continued for another minute or so between the jet and the approach control, and then Alexy turned his attention to the flight that had called out the spine-chilling Mayday alarm: "Clipper two one four . . . are you still on this frequency?"

There was no answer.

Clipper 214's last transmission was at 8:58 P.M., and now it was 9:05. Unknown to Alexy, the Clipper was a flaming funeral pyre for eighty-one persons. As the flames mounted higher and higher into the storm-filled skies, National Flight 16 scooted out of the area because flying over Elkton, Maryland, was no place to be this stormy night. That much was certain.

2

The Phenomenon of Lightning

Soon after this dreadful accident, the Civil Aeronautics Board, which at that time was in charge of air crash investigations if fatalities were involved and has since been replaced by the Department of Transportation Safety Board, made a preliminary report that turbulence appeared to be the culprit of the Elkton crash.

Despite the fact that scores of persons on the ground saw lightning envelop the Pan Am Clipper seconds before it burst into flames and that the pilots of the National jet also saw an enveloping flood of lightning around the Clipper which was just ahead and below them, the CAB made the preliminary guess that turbulence had torn the airliner apart. They may have been helped in this decision by two factors. First, there wasn't a pilot flying in the country who believed that lightning alone could cause a midair explosion, and with the exception of two lightning-suspected disasters, the facts seemed to bear out this contention.

Second, Donald Nordstrom, aeronautical engineer for the Boeing Company which built the 707 series, said he could find no lightning-strike marks on the fuel tanks or the plane's fuel venting system when he examined the wreckage at Bolling Field in Washington, D.C., and "to my knowledge there never have been any fires occurring as a result of such strikes."

But the experts and the so-called experts were wrong. Lightning did cause the explosion. It took the ultra-efficient CAB—undaunted by the mounting criticism leveled at the investigation by the industry—to trace the step by step events in the sky that night that led up to the explosion. In 1964, the United States talked about travel to the moon and atmospheric probes of Mars and Jupiter, but the country's technology was so far behind the times that on our own planet little was known of the great storms that created lightning and other natural hazards. Yet the jet age had no time to wait for technology to catch up, or it would never have got off the ground.

From the time the first passenger airliner was built, wing tips had been provided with lightning static wicks to bleed away electrical discharges. When the jets came along, some engineers thought that the thin knifelike tips of the swept wings would act as static bleeders. There is nothing like a fatal accident to change a way of thinking.

CAB investigators found that the Boeing 707 did not carry static wicks. But the DC-8 series did. Pan Am's Clipper blew up when hit, while the National DC-8, struck at almost precisely the same moment, did not. CAB therefore issued a directive that all jet aircraft carry wicks in the future, a decision made only after eighty-one persons had died. All this indicated that the behavior of jets under adverse weather conditions was unknown, and all the wind tunnel research, deliberate structural disintegration, and flight testing couldn't solve the peculiarities that seemed to be associated with thin-wing flying. Only a few weeks before the Elkton crash, Trans-Canada Airlines (now Air Canada) lost a DC-8 near Montreal with a record toll of 118 people and investigators were at a loss to find the cause. In fact, they never did find it officially.

Therefore, it was important to the industry and to commercial aviation around the world that the CAB find the cause of this Elkton crash. There were too many experts concerned mostly with protecting the image of the manufacturers and the airlines, and they were loudly critical of any suggestion that a mere lightning storm could turn a modern jet into scrap metal. Not when millions of dollars in advertising claimed otherwise.

Still, their criticism had some justification because lightning has been around for some two billion years and very little was known about it. What we call thunderstorms in reality should be called lightning storms, because the thunder is only the sound that the lightning makes as it arcs through the skies. Lightning *always* occurs in rain, something that wasn't known until just a few years ago.

Ironically, the first known lightning experiments in the United States took place close to where the Pan Am jet went down. Benjamin Franklin in 1752 attempted to confirm his theories as to the nature of electricity and made a number of experiments, the most picturesque of which was the famous one of utilizing a kite and flying it during stormy periods. The kite experiments helped prepare the way for Franklin's development of the lightning rod, which provided a means by which an electrical discharge could enter or leave the earth without having to pass through a nonconducting part of a structure, such as wood, brick, stone, or concrete.

An observer knows when he is dangerously close to a lightning bolt because he will both see and hear the discharge in the same second—a tremendous flash of brilliance and an ear-splitting explosion. The electrical discharge has broken away from the cloud above and has formed an ionized pathway to the earth. The charge is so tremendous that it heats to incandescence the molecules of the air in its path as well as any dust particles in the near vicinity. This heating of the pathway causes the violent expansion of the air around it, and this results in instantaneous compression of the surrounding air. The wave that is set up by this expansion and compression comes to the human eardrums as thunder.

The energy released by the average bolt of lightning lasts about half a second, and in the average storm there occur some ten to twenty strokes per second. Whether forked or zigzagged, each stroke releases a fantastic amount of energy, estimated at 100,000 kilowatts.

A single stroke is equivalent, in energy released, to fifty

gallons of gasoline. All this, of course, is not delivered to the ground. It is distributed along the entire length of the ionized path between the cloud and the ground. Yet the comparison indicates the awesome power of lightning and the dangerous ionized path of gas that it creates. If a flash of lightning should occur immediately in front of an airliner, so close that the craft flies into the ionized column of gas, it could be destroyed either by the explosion of fuel and air in its fuel tanks or by the destruction of metal and moving parts caught in the ionized zone.

This would be a freak accident indeed, but an entirely possible one. Witnesses to the Pan Am crash distinctly saw a blinding flash of lightning in the immediate vicinity of the jet, followed by the appearance of fire. There were some who recalled that back in the early days of crash investigation, such a freak accident did occur, but the Civil Aeronautics Board generalized the cause as a thunderstorm. Loss of control in turbulence seemed a reasonable assumption. That was that.

Let us have a look at the facts. On August 31, 1940, a Pennsylvania Central Airlines DC-3, with twenty-four passengers and crew aboard, was flying over Lovettsville, Virginia, during a severe electrical storm. Ground witnesses saw a tremendous flash directly in front of the aircraft which immediately plummeted to the ground. No one would believe that the electrical bolt had anything to do with the disaster, the first accident investigated by the newly inaugurated Civil Aeronautics Board.

When searchers dug up a battered alarm clock in the mass of blackened wreckage, it was thought they had found the logical answer: a bomb, triggered by the clock. FBI agent H. C. Crowley rushed the "find" to his Washington headquarters. Investigation of the clock turned up several important items. No scorching or explosive residue was found. Pieces of cardboard forced into folds of its case were found to be from its packing box. The alarm mechanism was set in such a fashion that it could not have triggered a bomb for almost seven hours after the crash.

But the cause had to be found, if the CAB was to survive.

The fact that the public in those days believed all airplanes eventually crashed was not a sufficiently valid excuse to write the crash off as an unexplained "act of God" incident. Back into the wreckage went the CAB experts and they started on a microscopic examination of every shred of metal, cloth, and, yes, even the passenger manifest. The latter, together with an envelope, had unmistakable charring on their surfaces. Then, on the carbon copy of the manifest, a small hole was noticed on line 14. Examination of this hole disclosed no evidence of charred fibers in the immediate vicinity, but further experiments, conducted in the Technical Laboratory of the FBI under the personal direction of J. Edgar Hoover, revealed that high-voltage electric sparks passed through paper similar to that of the passenger manifest could make a similar hole without any trace of charring. The conclusion was: "It is therefore possible that this hole could have been made by a bolt of lightning."

Meteorologists, studying the electricity of the atmosphere within ten miles of the earth's surface, seem to agree that the electrical field of the earth is constantly maintained by a phenomenon known as point discharge. This is a quiet and invisible process, and the build-up of the charge leaves the earth by the fastest means possible—that is, through sharp points and corners, like lightning rods, blades of grass, treetops, steeples, grains of sand, and so on. When a voltage differential of sufficient intensity takes place between the earth and the atmosphere above it, the discharge becomes strong enough to ionize the air and stir the ions to luminosity. This can be seen as St. Elmo's fire. It is a phenomenon worth remembering, for to an airliner it can be a warning of a lightning strike.

When thunderstorms wander over the face of the land, the constant process of the point discharge reverses because of the high-voltage differential between the cloud and the earth. The point discharge leaves the charged body of the clouds by sharp and clearly defined points for the area that provides the lesser voltage differential. Lightning bolts are the result.

Thus, when an aircraft flies through vapor-filled skies or in thunderstorms, the static electricity build-up created by the frictional forces of the flight leaves the metal surfaces through the sharply defined extremities, the thin wing tips. These are provided with trailing wicks to facilitate the discharge.

As this electricity bleeds away, it sets up an ionized pathway, which could intersect a flammable fuel and air mixture if the fuel tanks are accessible through a venting system.

The scientists who appeared before the CAB at the Elkton investigation clarified the subject by explaining that a lightning stroke begins when the air's resistance to the passage of electricity breaks down. At that precise millisecond, a faintly luminous stepped leader (a pulsing direction-finder for the coming lightning bolt) advances toward an area of opposite potential, which would be the earth in the case of a cloud-to-ground lightning bolt. The difference in the electrical potential between a cloud and the ground may be in the proportion of 100,000,000 volts to ten. This stepped leader advances toward the ground in a series of branching movements, forming the ionized pathway along which the discharge takes place. If this stepped leader approaches a flying aircraft, the intense electrical field induces streamers from the extremities of the aircraft. A contact is formed between the stepped leader and the aircraft streamers, completing the ionized channel and raising the potential of the aircraft to one hundred megavolts. This high potential instantly produces more streamers from the available points of the aircraft, and these streamers have sufficient energy to ignite fuel vapors.

Meanwhile, the stepped leader continues on from the aircraft's vicinity to another cloud, or to the ground, to complete the ionized channel for the electron avalanche that follows and that will heat the channel to 15,000 degrees Centigrade. This explanation of the birth and behavior of lightning would explain the mystery surrounding aircraft strikes and disasters.

The oldest recorded lightning strike on an aircraft took place way back on July 20, 1922, at 12:30 P.M., over the western foothills of the Rocky Mountains, not far from the city of Red Deer, Alberta. The plane in question was a single-seater De Havilland Four, which had been converted to forestry patrol work by the Alberta provincial government. The pilot, whose name has been lost in the intervening years, reported that he was flying between two giant clouds, probably of the cumulo-nimbus type. The pilot seemed to remember turning into one of the clouds, and he immediately had the sensation that his head was being thrown back. He momentarily blacked out. He came to his senses, feeling intermittent shocks coursing through his body, the strongest shocks being in the vicinity of his mouth and cheeks. He attempted to remove his microphone strap but was unable to move his fingers. He was fully conscious but paralyzed, and his DH-4 was in a gentle dive toward the ground.

The Royal Canadian Mounted Police, stationed at Rocky Mountain House beneath the towering clouds, were attracted to the sounds of the light plane. In those days any aircraft overhead brought everyone into the open to watch it fly by. What the Mounties saw on this occasion was unusual. The plane would roar and then dive. Next, it would nose up and stall. Then it would fall off, the engine would come to life, the plane would level out, and then it would begin to climb and go into a stall again. This went on and on as though there were no one flying the aircraft.

In reality the pilot and the DH-4 were at the mercy of a storm. Electric shocks were forcing the pilot's left hand to move the throttle erratically, while the right hand was moving the control stick back and forth. The unfortunate pilot was unable to control the jerking movements of his hands until the light aircraft was close to the ground. At this point he somehow regained control and landed on a level stretch of grass.

For the famous Mounties, who would walk for years in order to get their man, to have one drop out of the skies into their laps was a most unusual experience. Two troopers removed the jabbering pilot from the cockpit, while Indians gathered around to stare at the strange bird which had behaved like no eagle they had ever seen. The pilot was badly burned about the face, particularly in the vicinity of the right cheek, where he was accustomed to hold his microphone. His back, neck, and shoulders were painful when touched, and he complained of soreness when he moved. It was estimated by the police that the pilot must have been unconscious in the light aircraft fully one minute before the continuing electrical shocks transmitted their messages through his hands to the controls.

In the decades that followed this incident, hundreds of lightning strikes against all sorts of aircraft occurred, but reports were negligible until the first big passenger planes came into service. The first report on record in the U.S. concerned a DC-2's being "electrified" on August 18, 1937, during a routine flight near the edge of a thunderstorm. The pilot recalled that a giant spark of electricity jumped from one of the engine controls to the hand of the copilot. The gap was more than six inches. The pilot remembered a lightning flash at the time of the incident. The copilot was unhurt by the experience, but when the aircraft landed an examination of the fuselage revealed that a large hole had been burned in the left wing, near the body of the aircraft, and there were thirty burns on top of the fuselage and seventeen other burns on the wing.

On November 1, 1938, a Lockheed 14H2 was flying close to a storm when a lightning bolt discharged in its vicinity. The pilot was shocked by the blast, but more important, one of the engines stopped and had to be restarted. Then, on October 19, 1939, a giant Boeing B-314 flying boat was struck, and the damage to it was as follows: ten rivets burned out on the wing; nine rivets on the bottom of the hull; four burns on the fuselage, 100 feet of antenna lost; safety gap inside transmitter burned away; and attachment screws burned one-eighth inch in depth.

One of the most frightening experiences sustained by pilots and passengers during the earlier days of flying occurred on March 21, 1939, over western Arizona. The incident is a textbook picture of a lightning strike against an all-metal aircraft.

The DC-3 (the famous offspring of the earlier DC-2) cleared Burbank, California, for a routine first lap to Winslow in north-central Arizona. At the controls were Captain William H. Dowling and First Officer W. A. Jamison. It was midnight and the passengers were settled down to sleep as the aircraft climbed noisily through 3,000 feet of fog. The night was reasonably clear thereafter, until the flight arrived near Ash Fork, Arizona. At this point, the crew was aware that large cloud build-ups lay ahead, illuminated by the occasional flash of lightning. The aircraft entered these clouds east of Ash Fork, and the pilots were now flying on instruments at the assigned height of 11,000 feet.

A few minutes later, the pilots were startled by the appearance of St. Elmo's fire on the two propellers. It lasted for some twenty seconds, then their attention was drawn to the nose of the plane, where miniature lightning bolts, looking like forked lightning, began to discharge. Light rain was now falling. The cloud was thick and heavy. The temperature was thirty-two degrees, and a slight amount of ice was forming on the leading edge of the wings.

The clock on the instrument panel read 1:11 A.M.

All the essential ingredients were present, and then it happened. There was a blinding flash at the nose of the DC-3. There occurred an earsplitting roar. The plane lurched. Passengers screamed. Both pilots were blinded for at least a full minute, and the airliner flew erratically on its own with no hands at the controls. The air roughened, and little by little the vision of Captain Dowling and copilot Jamison returned sufficiently so that they could read the instrument panel. The first thing the pilots attempted was to report the incident by radio, but both transmitter and receiver were out of commission. The compass appeared to be working fine, however, and after clearing their heads they climbed to 13,000 feet to get out of some of the thick cloud. They then proceeded normally to Flagstaff, where the line of beacons was picked up (a most reassuring sight in those early days of cross-country flying), and thence to the nearest airport. The passengers were removed from the plane at Winslow to await another airliner.

An inspection revealed that the trailing antenna wire had been burned off at the plug; the tube containing the emergency trailing antenna was kinked so badly that it was impossible to release the antenna; the knife switch of the emergency antenna was broken; the tail cone was blown up and all rivets were gone for twenty-four inches forward of the taillight, which was hanging by its wires and, incidentally, still flashing. There was a foot on top of the nose of the plane on the left side where the metal was burned and blistered.

Since the nose and the tail were the only areas damaged, it was determined that the lightning bolt struck the nose and traveled along the wet outside skin of the fuselage, discharging from the outermost part of the tail assembly.

The report of this incident was widely circulated among pilots and served to bolster the belief that lightning bolts would cause only superficial damage to airplanes with no injury to crews or passengers. On March 25, 1940, another DC-3 was struck on the nose, and the pilot reported that it splashed fire in all directions. The trailing edge of the right aileron was burned and fused, but the aircraft proceeded normally to its next destination. Later, on May 1, 1940, still another DC-3 was struck on the nose with a blinding flash and a roar like a motor backfiring, but the crew had been forewarned by the appearance of St. Elmo's fire around the props which were throwing fiery streamers for fifteen feet. Sparks shot from the instrument panel and then came the flash. The crew members were blinded for many seconds and their arms and legs ached for some considerable time after the incident.

On April 13, 1939, at 10:00 P.M. while outbound from Chicago at a cruising altitude of 8,000 feet, another Douglas airliner entered sleet and heavy ice crystals. The searchlights on either side of the plane revealed ice accumulation. The radio reception became poor because of increasing static, and shortly thereafter the entire plane began to sparkle and glow with St. Elmo's fire. Electricity began to build up on the nose of the aircraft, and the pilot was forced to turn up the instrument panel lights and cockpit lights to cut down the terrific glare of the electrical fire, which played a nerve-shattering tattoo on the windshield.

"By this time, the sparks had increased to ribbons of flame, which I estimated at six to eight feet long and from four to six inches wide," the pilot wrote in his report. "This kept up for several minutes, and then there was a concussion like a shot gun, and a blinding flash of light enveloped the cockpit. For three minutes, I was stunned and unable to see. The automatic pilot was engaged. The bolt or the discharge seemed to enter the left side of the plane and pass toward the rear. After this happened, we were in comparatively clear air, and on arrival at Omaha we had the ship inspected. We found a burned tear in the right flipper on the trailing edge. The fabric had been split and the metal edge melted."

There were many instances of lightning strikes during those early years, which were reported to meaningless government agencies. Yet, even after the Civil Aeronautics Board was formed in 1939, pilots' reports of lightning strikes were secreted in the files of the operators and no statistics were kept on the subject. These effects of lightning strikes on aircraft were minimized to such an extent that pilots often boasted that physically lightning was absolutely harmless, that the only damaging effects were psychological. It was considered cowardly for a pilot to report a strike or any delay to his flight because of the proximity of lightning to his flight path.

To this very day, airliners take off below the base of ominous clouds, where lightning bolts are cascading to the earth, without so much as the slightest deviation in the flight path to avoid the situation. Planes repeatedly take off into storms, the pilots hoping that the odds against being struck will protect their craft from damage.

But today's jets appear to be more susceptible to strikes than their piston predecessors, perhaps because they are more vulnerable the faster they are flying. Scientists will not venture a guess as to why jets may attract lightning strikes more than piston-engine aircraft. Yet it is strange that, with a number of aircraft in a holding pattern, all within a few miles of each other as they were that night south of Philadelphia, the only planes to be struck were the big jets. Was it possible that jet engines created their own ionized paths with their tremendous heat and thus attracted lightning discharges from surrounding clouds? Only years of study could answer this perplexing question.

In October 1961, two airliners took off from Orlando for points north while a storm was raging at the south and east perimeters of the airport. The Delta DC-6B and an Eastern Electra plowed off into the lightning display with no strikes and a graceful climb into and out of the storm clouds. However, on Wednesday, March 13, 1963, under similar circumstances, a Tag Airlines De Havilland Dove, plying its regular short-haul route between Cleveland and Detroit, was struck. The pilot and three passengers escaped with a good scare, but the plane suffered seven holes in the fuselage and had its electrical equipment knocked out. Shortly afterward, a National Airlines Electra with forty-four passengers aboard, en route to New York from Newport News, Virginia, was struck by a bolt over Snow Hill, Maryland. Damage was minor, and only the plastic radar nose dome had to be replaced. On February 12, 1966, a TWA 707 jet was struck by lightning while making its approach to Fiumicino Airport at Rome. The radar in the nose was badly damaged, but passengers and crew were unhurt.

Lightning strikes, naturally, are not confined to commercial airliners. Military jets tangle with electrical storms from time to time, although all pilots are warned constantly to stay away from thunderstorms, even if it is necessary to turn completely around and fly back. The USAF pilot of an F-104 reported that, during his climb out, his high-speed jet was struck. There was a loud noise and a brilliant flash, occurring at the very moment that he had switched from his afterburner range to military power. His first thought was that the compressor had stalled, but all the instruments showed engine operation to be normal. However, radio communication was out of commission, power supplies were ruptured, the nose radome was holed on the topside, paint was chipped away on the nose, plugs were burned out, boost-pump circuit breakers were popped, and radio power-supply was burned out.

Air Force pilots are repeatedly warned to avoid the freezing level in a thunderstorm, if it becomes necessary to fly through one (most unlikely in peacetime operations, reads the bulletin). This is good advice. Lightning usually occurs in cloud layers where the temperatures are between fifteen and thirty-two degrees Fahrenheit. Air Force pilots are told to fly with cockpit illumination as bright as possible.

Pilots' handbooks claim the possibility of a strike can be minimized by reduction of speed. Earphones should be worn ahead of the ear to avoid possible deafness, and only one crew member should wear earphones during electrical storms. Lightning strikes, according to the CAB, are relatively frequent. They average one per 2,500 hours of operation for a piston aircraft; one every 3,800 hours for a turboprop; and one per 10,400 hours on a jet. They are, indeed, frequent; yet how many pilots in today's crowded airways shove their earphones forward from their ears to protect themselves from temporary deafness? How many wear sunglasses when flying in the vicinity of lightning? How many avoid the temperature levels where lightning could normally be generated? Few, if any.

When an airliner is caught in lightning it is unbelievable that its crew would remove their earphones. This is the moment when static is at its loudest, when hearing is difficult. Even if the plane is not stacked up in a holding pattern, the crew is listening to every shred of the continuous conversation that is going on, for it is essential to its own safety to know where other aircraft are located and to listen constantly to the weather reports in the area. On the night of the Elkton disaster, the busy controller at Philadelphia was up to his ears in calls and messages.

The crash of the ill-fated Clipper at Elkton reminded some people of the incident of a TWA Constellation that blew apart over Italy in 1959 while it was flying through an electrical storm. An explosion apparently occurred near a wing-tank fuel vent. A fuel vent is located near the end of each wing to equalize the pressure on the inside of the tanks with the pressure of the outside. It was recalled that after the Italian crash, the FAA had asked Lockheed, builders of the Connie, to investigate fuel venting in relationship to electrical discharge or lightning strikes. On May 31, 1963, after almost three years of study, Lockheed engineers sent the FAA a startling report.

"Yes," it said, "lightning can ignite an inflammable mixture spewing out of the vent. It would not be necessary for the lightning to actually strike the vent. It could strike anywhere on the aircraft and bleed off into the path of the flammable vapours. High-energy lightning strikes produced in the laboratory repeatedly produced ionized gas which readily passed through flame arresters of the type installed on jet airliners and moved such hot ionized gas right into the fuel tanks."

Despite this ominous report, nothing was done to change the position of the fuel vents on the big jets. The Civil Aeronautics Board issued a second report on the Maryland crash. This said that it believed the explosion might have stemmed from the ignition of combustible fumes at the fuel vent outlet by a lightning stroke, and again the airline officials in the United States did not accept the CAB's conclusion. Not by a long shot. Spokesmen for the industry before a government subcommittee on airline safety said they viewed the CAB's opinion as "presumptuous," and that it would be "extremely unusual for a lightning strike to explode fuel fumes."

Had everyone forgotten that in 1962 the British Air Ministry, following a comprehensive study of jet fuels, reported that lightning could ignite the flammable vapors in aircraft fuel tanks?

When the CAB undertook the Elkton crash investigation, Truman Finch, an old hand at such work, headed up the team. With him were William Hendrichs who, as hearing officer, would round up all the available witnesses and take their statements; Tim Chiu would probe the airliner's structure to learn where it had come apart; William Lenehan would retrieve and make a detailed analysis of the broken power plants; Billie Hopper would make an aircraft-systems analysis; Charles H. Leroy would probe all elements of the human factors in the case; Orion Patton would rebuild and read out the flight recorder, as he had done in many other recent jet crashes and near disasters; Bob Ruchich would investigate the Philadelphia air-traffic control procedures; Alan Brunstein would re-examine the weather; and Joe Silva would check the general airline operations and procedures of all those involved.

The unbelievers had to be shown a probable cause that would stand up in litigation, and the CAB plunged into one of its most intensive investigations to prove beyond any doubt that lightning caused the Elkton crash. They must prove that suggestions by the industry and the press that the cause might have been a bomb, or a structural breakup, or a rupture in the fuel systems, or an engine explosion, would not hold up against the lightning thesis.

The CAB, usually stunned if an investigation cost more than $10,000, would spend some $125,000 on this probe. This amount would not include the costs of the separate investigations by the FAA, the Boeing Company, the engine manufacturers Pratt and Whitney, and a host of aircraft suppliers from all over the country.

The first job was to round up and interrogate all the eyewitnesses. Most of these were residents of the outskirts of Elkton. "The sky lit up like a tremendous bolt of lightning . . . then there was a loud explosion, like thunder outside of my house . . . then we saw the plane burst into flames and fall apart," said Gerald Cornell.

Raymond Gregg was watching television when he was attracted to a brilliant flash in the sky that reminded him of dawn. With his wife Joan and their thirteen-year-old son, he rushed to the basement of his home as the fireball descended toward them. Parts of the wreckage struck the house, and major parts of the plane landed only 150 feet away. The family escaped unscathed.

Oliver Johnson lived a mile from the crash site. He saw a brilliant flash of lightning and, simultaneously, an orange glow appear in the sky. The glow trailed like a meteor, heading westward. It was wide and bright. He watched it develop into a brilliant fireball and descend to the ground. He drove to the scene and found burning debris on either side of the roadway. He parked his car and moved toward the fire, hoping to find some sign of life. What he found, instead, was a pilot's jacket in the mud. Woven on it was a Pan American wing insignia.

Joseph Dopirak saw the sky light up. He ran to his front door—just in time to see the falling jet "coming straight down with only one wing." It blew to smithereens when it struck the sodden ground, he said.

Arnold Turkheimer, of the Bronx, New York, was driving on the Maryland-Delaware Expressway when he saw a brilliant flash of lightning cross the road ahead of him. A glow in the sky appeared, becoming more brilliant as it traveled earthward.

Some people thought it was an atomic bomb. A skater, Jerry Greenwald, thought it was a bomb, until he could see it was an aircraft with one wing torn off. He was close enough to see bodies tumbling out of the plane.

George H. Lewis, an auto assembly-plant worker, and his wife Johanna said there was lightning at the time of the crash and for some time before it. "I had heard a jet going by," he said. "Then I saw and heard the lightning strike, and I looked out and saw the orange flames in the sky and I knew right away what had happened. The flames were coming out of the plane slowly, and the wind was pushing them back. It looked like a giant torpedo in the sky. It was traveling in a very short circle for a plane to make, turning to its right. The way it was turning, it looked like it would hit right here. It missed us, and I thought it had hit my neighbor's house, and I ran over to see if they were all right. The plane had actually crashed fifty yards from their house, but flaming wreckage was everywhere."

At the crash site the next day it was found that the jet had been literally smashed into unrecognizable junk. The flight recorder, built to withstand 100 Gs—units of gravitational force—had apparently been slapped with double that amount of force; its tape appeared to be hopelessly damaged.

State police began marking the bits and pieces that had once been living flesh. A dozen pathologists and undertakers began removing what was left of the bodies. It would take them until the end of the year to identify seventy-eight of them.

The CAB learned that the plane had flown a total of 14,609 hours—or more than 608 days of steady flying. It had survived a violent maneuver in a training flight over France, during which the right outboard engine had been ripped from its wing support, but this had been almost four years before—February 25, 1959, to be exact. Any suggestion that this particular Pan Am 707 might have been weakened in the violent mishap was discounted. It had been through four major overhauls since then.

There was, however, the question of metal fatigue. This aircraft was the oldest jet in service, and already the United States Air Force had found evidence of metal fatigue in similar jets which were being operated much less than commercial versions. One and a half billion dollars had been appropriated to have them beefed up by Boeing at Wichita, Kansas.

Metal fatigue had been experienced a decade earlier by the sleek British Comets, and the specter of other unexplained jet disasters loomed before the investigators. They were thinking of a 720B crash in the Everglades, the recent Trans-Canada DC-8 disaster at Montreal, a DC-8 death plunge off Portugal, a disastrous Comet dive off Bombay, and the involuntary plunges of a number of jets in the United States in the latter half of 1963. After several weeks of probing the wreckage and studying some of the meager scraps of information from other sources, the CAB reported that all evidence pointed to an in-flight fire, and to the apparent explosive separation of the outer portion of the left wing, which landed two miles east of the main wreckage site.

"These pieces apparently fell to the ground as free objects, minus sections of panel and skin from the fuel-tank area," the report said. "The bottom skin in the fuel-tank area was separated in a downward direction which indicated an explosive force. The location of the explosive force and the physical indications lead to a preliminary conclusion that there was a fuel-air mixture explosion in the left-wing fuel tanks. Accordingly, lightning appears to have been the culprit in the case."

To support their view, the CAB interviewed a total of 140 ground witnesses. Ninety-nine of these reported sighting an aircraft or a flaming object in the sky. Seventy-two saw lightning, and seven said that they saw lightning strike the aircraft. Three others saw a ball of fire at the fork or one end of the lightning stroke. Seventy-two witnesses said the ball of fire appeared concurrent with or immediately following the lightning stroke. Twenty-three saw an explosion in flight. Twenty-eight saw an explosion at impact with the ground. Never had there been so many witnesses to a nighttime storm disaster.

An additional twenty-eight witnesses saw objects fall from the aircraft in flight, and forty-eight persons described various portions of the aircraft which they had observed to be in flames. Since the wreckage was strewn over an area four miles long and one mile wide, the descriptions by the witnesses of the in-flight disintegration of the Clipper from the many locations where they stood in the storm helped the CAB to study the destruction with great care. Nearly six hundred pieces of wreckage were found outside the main impact area. These had to be gathered up and transported to a nearby building to be reassembled into a 707 mock-up, a costly and painstaking job.

The piece of wreckage farthest from the impact area was 19,600 feet away, and the nearest was 8,200 feet along a path 1,500 feet wide and two miles in length located some distance south of the main crater. Nearly all the remaining wreckage was scattered in a 600-foot circle around the crater. Some 1,440 feet from the crater, two of the engines were found with their attachment pylons and parts of the aircraft cowling. The other two engines were located in the main wreckage site. The examination of the wreckage in conjunction to its distribution disclosed multiple indications of lightning damage, fire, and disintegration in flight. Multiple lightning-strike marks were found on the left wing tip, and a large damage area extended along the trailing edge, or rear portion of the wing. The wing of the Boeing 707 was 130 feet 10 inches from tip to tip. The wing is more than an aerodynamic airfoil to provide the phenomenon of lift to keep the aircraft in the air. It is also a giant fuel tank with a capacity of 17,406 gallons, distributed throughout the entire width of the wing in a series of tanks that ranges in size from a small surge tank on the wing tips to a giant center tank beneath the main fuselage. The center tank holds as much as a gasoline-delivery tank truck.

From the tips of the wings to this center tank, the distribution of gas tanks is as follows: Vent Surge Tank; Reserve Tank (434 gallons); Main Tank No. 1 (2,333 gallons); Main Tank No. 2 (2,283 gallons); Center Tank (7,306 gallons). A ventilating system connects all fuel tanks, and there are vent outlets near the wing tips to expel accumulations of fuel-and-air mixtures. These permit fuel to be expelled when it is sloshing in the tanks, or when too much is loaded with the resulting increased pressure in the fuel tanks.

The lightning-strike marks were found near the tip of the left wing in the vicinity of the Left Reserve fuel tank. The damaged area extended from the trailing edge of the wing to three feet eight inches from the forward edge. Within this area there were numerous spots where the metal surface and the rivets showed evidence of melting. The largest single clue was an irregular-shaped hole about one and a half inches in diameter, with evidence of metal fusing around its edges. Another smaller burned hole was found nearby, and near it four rivet heads were burned off. Numerous small lustrous craters were found in the metal wing skin. Paint had been charred, and specks of fused metal were found on the paint. The bottom of the Left Surge Tank was fractured, and the tank was bulged outward, showing the effects of an internal explosion. The most lightning damage was found only eleven inches from the vent outlet pipe.

Pan American conducted a flight test in a 707 to determine if fuel would discharge through the tank-vent-system air inlet, but at no time was there any visible discharge. There was evidence, however, that fuel entered the vent system, collected in the surge tanks, and then returned to the proper fuel tanks. This test assisted the CAB in reaching an obvious conclusion.

It was determined finally and officially that explosions from a massive lightning strike occurred first in the Left Reserve Tank and spread to the Center Tank, and the Right Reserve Tank and these multiple explosions occurred when the first explosion structurally disrupted the other tanks. This caused the disintegration of the entire left outer wing, and there was an immediate loss of control. Boeing 707s later experienced outer wing destruction by fire and by in-flight collision out of San Francisco and over Connecticut, but in both these cases the pilots managed to maintain control. In the Elkton case, the loss of control was almost immediate. Other explosions of fuel continued, scattering broken human bodies and the aircraft itself in bits and pieces over the wet countryside.

When the "probable cause" was made known, the Federal Aviation Agency at once alerted all pilots and air-traffic controllers of the dangers of lightning and the need to avoid it. Air carriers were asked to install static dischargers on all aircraft using turbine fuels.

Two airworthiness directives were sent to Boeing. One required modifications to the fuel-tank access-door bondings, and the other required an overlay to be made on the metal skin of the surge tanks for improved protection against lightning penetration.

Meanwhile, the Flight Standards Service of the FAA formed a technical committee on lightning protection for aircraft fuel systems, composed of representatives of the FAA, the CAB, the NASA, the United States Weather Bureau, the United States Air Force, and the United States Navy. Lightning-strike data from all available sources in the country were called for. A contract was signed with the Atlantic Research Corporation to evaluate flame-arrester designs, as well as studies and tests of protection for vent systems. Another contract was awarded to a research institute for investigation of internal arcing in present wing-tank construction with a view to eliminating such electrical arcing through the system of tanks. A study program was established to evaluate the safety of such jet fuels as JP4, a kerosene and gasoline mixture, some of which was in the Elkton 707. The FAA also asked for an immediate re-evaluation of all aircraft in service with regard to lightning-protection features, for they hoped to force such features on all turbine aircraft.

It took the Elkton disaster to bring all these things about. It should have made thunderstorm penetrations illegal. But it didn't and the disasters continued.

3

The Danger That Radar Cannot See

Lightning is only one ingredient of a thunderstorm that effects the safe operation of a jet airliner. Turbulence is another. Wind gusting and wind shears (a line between two oppositely moving windstreams) are equally dangerous and so too is ice and hail.

A thunderstorm manufactures energy of fantastic force within the lovely cumulus castles of a summer day or in the storms that are created by the clashing of cold air streams from the Arctic and the hot humid rivers of air from the tropics. All weather is created by the collision of these air masses, and the degree of violence is most often associated with the speed at which these hot and cold air streams meet. The swifter the clash the more eruptive the heavens.

Pilots who have flown both would far rather risk their lives in a tropical hurricane than fly into the cells of a mature thunderstorm over the Kansas Plains. The wind forces defy imagination. It wasn't until the arrival of jet airliners that the forces present at all levels of the atmosphere were experienced, and usually with disastrous results, proving again and again that our knowledge of thunderstorms is in dire need of improvement. And it seems that as we build aircraft to fly higher and higher, the known problems increase in intensity instead of decreasing and new problems are added to the old, as "seat of pants" experience instead of technology sets the rules.

There is no such thing as "all weather flying" and there never will be. Severe storms are lethal to both jets and supersonics, and recent high-flying experiments over Puerto Rico by the United States Air Force have shown that severe storms are found as high as 120,000 feet, a height that cannot be overflown in this century by commercial aircraft.

Thunderstorms have been ravaging the earth since the dawn of time. They occur someplace in the United States every day in the year, and weathermen studying the reports of the Tyros satellite have estimated as many as 1,800 thunderstorms occur daily around the world. With so many storms of this kind and any one of them a danger to commercial aviation, you would think there would be an international commission to study the problems and effects of severe storms on world travel. But there is nothing of the kind. Most airline research money for atmospheric study is being dumped into turbulence that cannot be seen—the kind of turbulence that has caused no deaths to date. The reason for this research will be revealed later.

Meanwhile, airplanes fall victims to thunderstorms from Calcutta, to Tokyo, from London to Cincinnati, year after year after year. Some 90 percent of all U.S. commercial air disasters happen in bad weather associated with thunderstorms. More than eight hundred fatal general aviation airplane crashes in 1967 (the latest statistics available) occurred in storms or other bad weather. Not only that, airliners do not have to be within the thunderstorms to be affected by the storm's lethal tentacles. So violent are the motions of air within cumulonimbus clouds that sympathetic violence is generated miles away from the storm center. This sympathetic violence is not generally explored and is rarely researched, but it is very dangerous to jets.

Every jet accident is different, and not just because of the immediate circumstances which involve weather and flying procedures, air traffic problems, and crew capability. Without exception every jet mishap has set new rules of flying and has demonstrated on an ever-increasing scale that the United States and other countries are woefully behind in the aeronautical technology necessary to keep abreast of increased jet flying. All this means that as more jets are brought into service and the more flights there are, the probability gap of disaster is narrowed dramatically. Never has this been more clearly demonstrated than in the crash of a Braniff airliner near Falls City, Nebraska, on August 6, 1966. The full impact of the investigation never reached the industry until mid-1968 and the public was not fully informed as to the real meaning of the investigation's conclusions.

In simple terms, the National Transportation Safety Board discovered that modern jet airliners, whether built in the United States or elsewhere, are not built strong enough to meet all flying conditions that are encountered today.

This discovery, at the present state of aeronautical art, was received with little enthusiasm. Behind closed doors it was feared that if government agencies and airframe manufacturers didn't face up to the problem, multimillion-dollar court actions might eventually force a change in jet construction.

It all began as a routine flight and a routine crash investigation. But as the probe continued by the National Transportation Safety Board, the manufacturers of the aircraft, and by the Air Line Pilots Association, future disasters appeared certain and the specter was never dispelled by the findings.

The flight was Braniff Airways, Inc., Flight 250, operating on a daily milk-run from New Orleans to Minneapolis with stops at Fort Smith, Tulsa, Kansas City, and Omaha. No untoward incident was reported from New Orleans to Kansas City, but after the aircraft departed the Kansas City Municipal Airport on the night of August 6, 1966, at precisely 10:55 P.M., things began to happen.

The aircraft was built by the British Aircraft Corporation. Known as a BAC-111, it was in wide service in the United States on a number of well-known carriers. The aircraft was very popular with passengers and crew members because it was fast and quiet and could be operated from the shorter runways of small city airports. Two Rolls-Royce by-pass engines were mounted on pods suspended at the rear.

Captain Donald G. Pauly, forty-seven, a veteran pilot with more than 20,000 hours to his credit, had been flying BAC-

111s for some fourteen months and liked them. He had satisfactorily completed his semiannual proficiency tests and the FAA inspector had noted in his report that Pauly's work was "very good." He had amassed 549 hours in the BAC-111 of which almost half was in nighttime flying. During 1966, he had flown a total of twenty-five hours on instruments, an indication that flying the mid-central routes meant many days of cloudy and rainy weather.

His copilot was First Officer James A. Hilliker, thirty-nine, who started out as a baggage handler and then became an aircraft engineer in 1955. He had become a commercial pilot in 1955 and since then had chalked up 685 hours in the BAC-111. Few pilots knew this aircraft better than Hilliker.

Back in the passenger compartment two stewardesses in the mod Braniff regalia were on duty for the light load of thirty-eight passengers. Seventeen minutes from Kansas City all would be dead.

Before the takeoff Captain Pauly had been concerned about the weather along his route. He discussed this with another captain who told him that on his route there was a "solid line with very intense thunderstorms with continuous lightning and no apparent breaks, as long and mean a one as I'd seen in a long time and I didn't feel the radar reports gave a true picture of the intensity."

For some reason this information was not relayed to the Braniff dispatcher. Until recently the decision to operate a flight had always been left up to the captain-in-command but, following a number of jet disasters in bad weather, the rules were changed so that both the captain and the company's flight dispatcher at the airport shared the responsibility for flying operations and therefore the blame. In this case, despite the fact that the captain showed concern about the weather even before his departure from New Orleans, the storms along his intended route northwestward from Kansas City appear to have been underrated by the Braniff personnel responsible for analyzing the weather and dispatching the flight. Captain Pauly did not obtain a formal weather briefing before leaving the Kansas City airport, and it is logical to assume that he fell into the trap of utter dependence upon his weather radar to help him.

Braniff flying procedures, like those of some other airlines, prohibit the dispatch of an aircraft into such weather. The Flight Operations Manual states that if detouring a solid line of storms is not practicable, flights will be held on the ground until the line has passed, has dissipated, or can be circumnavigated. The company, however, had forecast only scattered thunderstorms even though the Weather Bureau had issued a Severe Weather Bulletin (a copy of which was found in the wreckage) and had forecast severe thunderstorms and numerous cumulo-nimbus with tops up to 50,000 feet.

The dispatcher later stated that he would not hesitate to ground an aircraft, but that the weather situation on the night of August 6 did not, in his opinion, warrant such action. The Safety Board believes that Flight 250 was given the go-ahead in good faith, but that the faith was founded on inaccurate analysis of the weather over the Midwest that night—an indication of the confusion present too often in commercial operations. Inconsistencies in forecasting, as to degrees of seriousness of weather, may have lured worried pilots into taking chances they would not otherwise have taken.

The graceful aircraft zoomed off the Kansas City airport (often regarded as the most dangerous airport in the country because of its proximity to high-rise buildings) at precisely 10:55 P.M., with an instrument clearance to Omaha along Jet Route 41 at a cruising level of 20,000 feet. Before climbing to this height, however, the pilot was advised that because of cross traffic above him, he was to restrict his height to 5,000 feet until he received the signal to climb. Twelve miles north of Kansas City, the airport transferred the flight to the local Air Route Traffic Control Center and, when radar contact was confirmed, the pilots were instructed to climb up to the assigned height. But Captain Pauly and First Officer Hilliker with their radar could see ahead of them a wall of precipitation with numerous cells of thunderstorms and they knew that to collide with such a line was almost sure death.

And also they could see, as could the passengers, a band of heavy lightning. As they quickly reached 6,000 feet in their climb, they changed their minds about their assigned altitude because of the danger of flying into a series of storm cells. They asked to return to 5,000 feet in the hope they could keep under the storms ahead. Though they didn't know it, two tornadoes were lurking along that line. Apparently their small radar screen did not disclose the funnels.

There was also something else they didn't know—circumstances that had just occurred and about which neither their flight nor other flights were informed, showing the lack of coordination between companies and government agencies about safe flight operations even today. Another Braniff flight (255) had delayed its takeoff from St. Louis because of the bad weather at Omaha, and Braniff Flight 234 en route from St. Louis to Omaha had diverted to Kansas City after the pilot saw the storm ahead and elected to stay clear of it. Not only Braniff dispatchers but government traffic controllers as well knew these two important facts, but they failed to inform Captain Pauly and his First Officer. Negligence such as this usually ends up in multimillion-dollar litigations.

At four minutes after eleven o'clock, following a short conversation in the cockpit about the possibility of flying through an opening in the clouds, Flight 250 requested from Air Route Traffic Control a change in course, to the left if possible, to take advantage of the hole in the clouds, visible only on the radar in the cockpit. Two minutes later the Chicago Air Route Traffic Control called the flight and informed it that ". . . the line appears pretty solid all the way from west of Pawnee to Des Moines."

For the next three minutes the pilots discussed a deviation to Pawnee. The discussion ended with ". . . we're not that far away from it. Pawnee is a hundred and twelve four if you want it," the First Officer was saying to the captain, providing him the radio frequency of the Pawnee radio range. Their conversations were etched in the cockpit voice recorder.

The captain gave no indication that he intended to swing toward Pawnee as his copilot suggested, because he was apparently planning to penetrate the storm through the hole that had shown on his radar. The aircraft continued toward the wall of the storm and, somewhere about five or ten miles south of the nearest precipitation, the flight disappeared from the radarscope of the Chicago Center.

At 11:11 P.M., the captain told his copilot to ease the power back in anticipation of the jet's entering a known zone of turbulence. At this point the engines would sound much softer as the aircraft slowed down to 270 knots, the storm-penetration speed of the BAC-111.

Nine seconds after this command a strange noise enveloped the aircraft. It was not thunder. The noise increased to a constant level and could best be described as a "rushing air" sound. Eight seconds later, another unidentifiable sound on the cockpit voice recorder could be heard; following this came an electronic flutter followed by the raucous sound of four klaxon horns that warn the cockpit of a stall. The recording ended exactly twenty-eight seconds after the easing of the power to penetration speed.

The flight was dead. But what could have happened? The aircraft was at least five miles from the leading edge of the cold front associated with the line of thunderstorms. As it was flying at the relatively low altitude of 5,000 feet, could it have tangled with a tornado? It would take intensive investigation to learn the cause of this disaster.

The investigation had hardly got underway when twelve days later, another Braniff BAC-111 left Kansas City as Flight 233 and flew into almost identical weather conditions with turbulence so extreme that it was only good luck that saved it from a similar fate. This seemed to indicate that pilots were placing too much faith in radar and not enough on eyesight and Severe Storm Warning information.

As in the case of the Elkton crash, the Department of Transportation Safety investigators found many witnesses to the disaster even though it was a stormy night and late at that. But the Severe Storm Warning for the area had been well broadcast during the evening and hundreds of residents along the Nebraska-Missouri border were outside of their houses watching for signs of tornadoes. People in this area are familiar with violent summer thunderstorms and tornadoes, and watching severe weather is almost a way of life. Two funnel clouds were seen by witnesses eight minutes after the crash. Altogether some three hundred persons on the ground in the vicinity of the accident were interviewed and, according to where they made their observations from, the storm appeared different but always dangerous.

All witnesses agreed that the unlucky flight never reached the main line of storm clouds and that at the moment of the crash there was no cloud-to-ground lightning in its vicinity. Persons closer to the accident thought the plane had just entered some thin cloud before they perceived a mushroom of flame and the descent of burning wreckage to the ground. Persons at a greater distance thought the aircraft was above the lower clouds and more or less in the clear.

It was a case of too many clues. While the bits and pieces of the plane were being assembled for evidence of a structural failure, investigators fanned out to talk with other pilots flying that night in the vicinity of Omaha and Kansas City. A number of witnesses were located and interrogated. Four airliners had been in the area within an hour of the accident time. One was a Convair 440 that left Omaha for St. Joseph, Missouri, at 10:36 P.M. under radar control, with an assigned cruising altitude of only 3,000 feet, which would keep it under the base clouds of the thunderstorms. This was done in the mistaken belief that it is less dangerous to fly beneath a storm at low altitudes than it is to fly at higher altitudes. Five miles south of Omaha airport, the captain of this flight could see heavy shower activity straight ahead of him, and he asked and got a change of heading that would keep him farther east of the trouble. Under radar and by using the lightning flashes to outline the clouds in his flight path—which most airline passengers would find rather discomforting—the flight got through the storm with no more than moderate turbulence.

During this flight, the radar showed many large thunderstorm cells and a squall line approximately thirty miles long extending from Shenandoah, Iowa, to the northeast. More large cells extended south and west. Thunderstorms were everywhere that night as had been forecast. This captain said the cloud bases were between 3,500 and 4,000 feet, and he was just beneath them while lightning was almost continuous.

About twelve minutes out of Omaha the cloud base lowered and the flight went on instruments; the reader can only sympathize with the plight of the passengers as the plane entered moderate "plus" turbulence. Three minutes later, the pilot took the plane up to 5,000 feet. At this time the turbulence intensified and gusts were so severe that pillows and blankets were thrown out of their overhead racks. At 11:00 P.M. and twenty-five miles northwest of St. Joseph this flight broke into the clear.

The flight recorder of this aircraft was reviewed, and it showed that turbulence started a scant three minutes after takeoff from Omaha and varied for the rest of the flight from "moderate to severe turbulence." The most significant ups and downs were performed when the flight was eighteen miles east of the Braniff accident site.

It is significant that the lower speed Convair aircraft got through this storm while the higher speed jet failed to make it—one more of the all too prevalent indications that jets are vulnerable both in thunderstorms and in the vicinity of severe weather.

But there were other jets aloft that night. Braniff Flight 255, mentioned earlier, took off from Omaha at 10:55 P.M., but the captain noting a squall line on his route turned east and climbed to 17,000 feet in his BAC-111. After flying some forty miles, the pilot radioed Chicago and said he had found a hole in the cloud by radar and started through on a southerly heading. The plane was brought down to 7,000 feet and encountered light turbulence in rainfall for twenty to thirty miles. But when the flight left the clouds, the heaviest turbulence was encountered.

This jet was twenty-nine miles east of the doomed BAC-111, and the crew was in radio contact with Flight 250 until just prior to the accident. The flight recorder of Flight 255 showed that the airliner received turbulent buffeting four minutes after takeoff. It continued for a full sixteen minutes, and the greatest variations in gusting occurred when the craft was at 7,000 feet and due east of the accident site.

Meanwhile at 10:03 P.M. another Braniff BAC-111 departed St. Louis for Des Moines and was cleared to fly at 24,000 feet. But radar and visual observations convinced the captain of this flight that the storm ahead was lethal. He could find no breaks in the line and he diverted his flight to Kansas City. He may have been the smartest captain flying the area that night.

It was abundantly clear to the Safety Board that dangerous storms existed that night, but they were faced with the nagging possibility that the flight never reached the line of the intense activity. Searching through the wreckage they found that the throttles of the aircraft had been set for turbulence penetration, a much lower speed than cruising speed. This fact, supported by the evidence of the voice recorder's "ease power back," led the investigators to assume the aircraft had reached turbulence, or was just about to reach turbulence, and that the crew was prepared. Still, the aircraft was five miles from the storm line. So the Weather Bureau was called upon to make a special study of the storm situation that night and to be ready to support its conclusions. The Safety Board also called upon an independent meteorologist to make a separate survey of the conditions.

The wreckage of the aircraft was found scattered over a square mile of rolling countryside about seven and a half miles northeast of Falls City, Nebraska. The right wing had broken off in the air and the spilled kerosene erupted into fire. Of the 12,000 gallons that were in the wings of the jet at takeoff, only 9,000 remained when the disintegration of the aircraft took place. The wing fell half a mile from the main body of the wreckage. And most of the tail assembly was located 2,752 feet from the site, indicating that both wing and tail had been wrenched off in flight. There was no evidence of hail or lightning strike, nor evidence of static discharge.

Although most of the aircraft was badly burned with the exception of the cockpit, and all passengers had died of traumatic injuries when they struck the ground, there were enough clues in the cockpit and in the broken tail and wing pieces to reconstruct the disaster. While flying in a straight and level altitude, the jet was suddenly subjected to forces that caused it to respond violently, accelerating upward and rolling to the left. The automatic pilot was engaged and working. At this second the right tailplane—the horizontal section—and the tailfin snapped. The aircraft pitched nose down for a few seconds and then the right wing snapped, unable to bear the strain. The total time required for this sequence of death was between one and two seconds, an indication of the quick and powerful force of the gust. To better understand this violent progression of weather it is necessary to study the vast movement of air over the entire Central States that day. Scattered moderate rain showers and thunderstorms through Wisconsin, Minnesota, the Dakotas, and northwestern Nebraska had been forecast for that August 6 by the U.S. Weather Bureau. A cold front was slicing southward from Canada running from northeastern Iowa to the southeast corner of Nebraska, Central Kansas, and into the Oklahoma and northern Texas panhandles. There was to be some decrease in rain shower activity during the night, but thunderstorms were expected to remain along the leading edge of the cold front, and the Weather Bureau's Jet Level Forecast, in effect from midafternoon until five the next morning, indicated that moderate to greater-than-moderate turbulence would exist in the vicinity of all cumulo-nimbus clouds. The tops of these clouds were forecast to 41,000 feet, which meant that all levels of jet flying contained dangerous ingredients. And all this nasty activity was to move to the vicinity of Salina, Kansas, St. Joseph, Missouri, and northeastward to Dubuque, Iowa, by dawn the next day. Turbulence also existed in the lower atmosphere; this was borne out by the ascension of a weather balloon at Omaha at 7:16 P.M. which disclosed rough air up to 10,000 feet. Radar reports from Omaha, Topeka, and Des Moines all showed weather echoes (thunderstorm cells and intense precipitation and hail) over the same general area. A Severe Weather

Warning also bracketed the area, leaving no doubt what was going on in the atmosphere.

A pilot flying into such an area in today's jet aircraft depends upon his weather radar to steer him along the levels where the least possible turbulence exists. Since radar is merely an echo from something solid enough to give it an echo—such as raindrops that have a nucleus of dust—the pilot associates turbulence with rain. And it's true that rain creates turbulence because in its descent it carries the air along with it, setting up vertical wind shears. In thunderstorms the densest rain is found in the cells that are hidden in the interior of the great billowing cumulus clouds which are constantly changing as the air currents within them increase in velocity and height. In the very center of these storms there can be one or more cells of activity, which the pilot locates by radar echoes from the densest precipitation surrounding the cells. On his little scope they appear as circles and are often called "Cheerios." Clear-cut circles usually warn of severe wind shears. A hooked finger from the cell's perimeter will indicate a tornado or a hail cloud.

The pilot uses his radar to fly in relatively rain-clear areas or between the cells of energy. That this can be a dangerous and a lethal trap was not discovered until the BAC-111 disaster. Captain Pauly was no amateur at airborne radar manipulation; he searched the line of storm ahead and on either side, looking for a break in the precipitation. He was at a very low altitude for a jet, though this altitude was not low for small commercial or general aircraft. His radar showed clear space ahead of him, and he steered the BAC-111 into it.

But radar cannot "see" a rolling scud cloud or squall line which is common to the movement of such storms. From the ground, such menacing clouds herald the approach of the storm. They are black twisting vaporous clouds that hug the ground and are torn and twisted by the gale winds that accompany them. Such a rolling line is usually several miles ahead of the main storm and is created by the collision of winds that are rushing from above the ground to replace the winds that are being slammed downward inside the thunderstorm cells. These winds angle upward toward the thunderstorm, usually into the base of the clouds, and as a result they collide with the violent downdrafts of the winds that are dragged in by the precipitation within the cells.

As a result of this collision, a rolling motion is generated in a clockwise movement like a steamroller going over the surface of the earth. This rolling air drags more cloud with it and gathers up dust and debris and rolls ahead of the storm as a warning of the fury just behind.

Both the U.S. Weather Bureau and independent weather experts agree that the roll cloud that night extended downward for 5,000 feet below the base of the storm. The cold air along the ground traveling toward the storm was suddenly caught and dragged upward, where it collided with sixty-mile-an-hour winds heading southward and a little downward. Together they formed clockwise vortexes. The wind shears in this mad scramble were severe.

After having determined the motion of air in the storm the investigators were still perplexed as to what had caused the structural failure of the tailpiece and the first thing they could think of was the design of the BAC-111 tail in the shape of a "T." The Douglas DC-9 was built the same way. They wondered if the shape of the "T"-tail caused its failure when winds slammed at it from varying angles and possibly upset its flight characteristics and snapped it like a kite cross-piece.

Exhaustive tests were conducted by the Department of Transportation Safety experts and by the British Aircraft Corporation on the structural efficiency of the T-tail. The National Aeronautics and Space Administration was also asked to review and assess the aerodynamic design of the BAC-111 tail. All three agencies worked together trying to solve the mystery—a solution upon which most future aircraft designs might well depend. It was a critical period for the jets which had had disturbing critical periods before this. Designers pored over the aerodynamic features of new jet-age fin and tail surfaces. NASA conducted wind tunnel tests and subjected BAC-111 tailpieces to tremendous gusts from every conceivable direction and a combination of directions and then simulated the conditions that existed that night in the flight path of the ill-fated jet.

The result: strength of the T-tail and stability of the design were satisfactory and similar to other jets made world-wide. The structural design of the BAC-111 was found to be similar to other commercial jets, while the British method of testing was actually of a higher standard than was generally required for satisfactory performance.

The investigators on both sides of the Atlantic were reminded that when the FAA monitored the certification of the BAC-111 for use in America, special attention had been given to the T-tail, with particular emphasis on its stall characteristics. This was more intense than usual because the first BAC-111 stalled and crashed during a test flight near Hurn, England, killing some of England's best aerodynamists.

The FAA issued a report on the jet: "It is concluded that the BAC-111 is constructed and has been tested with the latest state of the art. The testing of structure, loads and fatigue have been very extensive and beyond the normal requirements. Environmental testing, with the exception of icing tests, have been more extensive than those required on like United States aircraft. It is believed that adequate corrective and preventative action has been taken in the design and systems to preclude similar problems as occurred during the developmental accident."

On April 15, 1965, the BAC-111 was certified to fly in the United States, and immediately Mohawk Airlines and American Airlines put them into passenger service. Until the Braniff disaster, and an accident that resulted in a rear-end fire in a Mohawk airliner which crashed with a heavy loss of life, the performance of the BAC-111 had been generally pretty good.

But despite the tests and the certification of design and structural qualifications, it was believed that the roll cloud conditions and the violent wind shears of that particular night had somehow affected the tailpiece, which in turn put loads on the right wing that made it crumple split seconds later.

At this point in the probe, the investigators were fearing they would come up with a cause that no one was going to like. The BAC-111 was strong and well built and met the standards of the United States, like all jets in service. But was the BAC-111 strong enough for storms, storms that seem to treat jets differently from other aircraft? Was any jet flying today strong enough to collide with storm conditions that frequent the U.S. airspace almost every day of the week, year in and year out?

Back to research and further testing went the probers, and herein lies a tale of disaster yet to come. It was found that many modern jets are not built strong enough to withstand wind forces associated with thunderstorms, wind forces that cannot be seen by either ground-based radar or airborne radar.

This conclusion shook the industry to its roots. With billions of flying miles at these relatively low altitudes where planes have been flying steadily since the first machine was invented, it seemed unbelievable that the Safety Board could have reached such a conclusion at this stage of the game. But the Board quickly pointed out that the situation was more critical today than ever before because of the jets whose thin wings and tails were built for speed and apparently for good weather.

It was determined that the gust velocity that caved in the BAC-111 was in the nature of 140 foot-seconds, which means it was a pulsating burst of wind moving 140 feet in one second.

And 140-foot-second gusts were admittedly out of the "limits of measured experience," although horizontal gusts had been measured in two Weather Bureau tests as high as 208 feet per second and 175 feet per second, respectively.

United States- and foreign-built jet aircraft are constructed to a design criterion that allows for maximum gusts up to 65 foot-seconds. This prompted the Safety Board to say: "If any good is to be derived from this accident it must take the form of increased knowledge relating to design and operation of aircraft in turbulent atmospheric conditions: of the nature of turbulence which may be expected, especially at lower levels; of the proper operational procedures to be followed if such turbulence must be penetrated; and of the forces and accelerations which may be produced on an aircraft by that turbulence."

4

Clear Air Turbulence

At the end of the first decade of jet flying what has been learned?

A panel of "million-miler" pilots and airline officials discussing this very matter before the Aviation/Space Writers Association in May 1968 agreed unanimously that one thing had been learned: airline safety had to be improved in the next decade, and improved dramatically or there would be a public revolt.

Jet service began in 1959 and since then an incredible eighty-nine jets have been lost and more than 400 have been involved in serious accidents (there was only one passenger death that first year). By January 1969, these accidents claimed 3,386 lives while 7,000 others died in crashes of piston and turboprop aircraft. The number of fatal jet accidents doubled between 1966 and 1967 and in 1968, regarded as a "good year" by the airlines, nine big jets went down and 372 persons died. In a survey of fifty-nine catastrophic crashes made by the International Civil Aviation Organization (ICAO) more than half of the jets involved were Boeing 707-720 series.

At the end of this jet decade more than 3,500 jets were flying in a world commercial fleet of 7,344 aircraft. Fifty thousand pilots were needed to fly these aircraft (with 12,000 more pilots being added to the jet fleet from the pistons and turboprops) because in addition to the crew, there were backup crews and all were working shorter hours. By 1975, it was estimated by the Air Line Pilots Association there would be a total of 90,000 commercial airline pilots needed to meet the expanding business.

Obviously by that time all landings and takeoffs will be fully automated, which ironically means that pilots will have to keep more than ever on their flying toes because they will be living daily with "hands off" operations. At any given time in a flight, a pilot may be required to take over controls which they haven't used for some time, yet they must be able to handle quickly and efficiently any emergency with infinite flying skill. All this involves an entirely new concept in training.

With so many aircraft and people taking to the skies, safety becomes number one on any airline agenda. No longer can the operators hide behind the idiotic comparison of the "passenger mile" to other forms of transportation. Since the end of World War II, less than 1,200 have died as passengers in railway accidents in the U.S. while deaths in commercial air carriers have totaled over 6,000 and in General Aviation (which also moves passengers) more than 21,000.

During December 1968 and January 1969, there were nine crashes involving U.S. commercial aircraft with a loss of 206 persons. Altogether 350 persons died in U.S. air carriers during the year 1968, a sordid record that was the second highest in history. In one 60-day period, ten airliners were involved in fatal crashes in the United States, Europe and Latin America.

What a record!

Today, the industry hopes to hold the crash rate to one accident in every 300,000 hours flown and by 1975 to hold the record to one crash in every 700,000 hours. Paradoxically, the loss of life will be higher in 1975 because bigger planes will carry hundreds more passengers and the dollar investment in each disaster will jump from six million to more than twenty million, and with supersonics up to sixty million. Litigation will skyrocket. (Settlements for the next-of-kin for American citizens killed in the BOAC Mount Fuji crash in Japan amounted to over $13,000,000.) By the end of the century, two billion people will be flying each year. No wonder that safety is the paramount concern of the airlines and yet how little is being done about it, despite lip service to the contrary.

When, as late as 1968, a group of representatives of the industry were asked where the safety emphasis was being placed, the answer was startling to say the least . . . on Clear Air Turbulence.

Why Clear Air Turbulence, since CAT, as it is called, has never caused a single disaster (although there have been some close shaves)? The reason, it seems, is that CAT has caused widespread structural damage to jet airliners and has knocked many of them out of service for lengthy periods of time. This in turn has played havoc with revenues. Repairs to jets from damage caused by Clear Air Turbulence have been estimated as high as $400,000,000. As to spending their research dollars on the kind of turbulence that causes crashes and heavy loss of life, the industry people seem to think that this will be solved by the weather experts augmented by more sophisticated cross-country radar surveillance systems. Perhaps they are right.

Clear Air Turbulence can best be described as the unseen wind forces that are created by the tidal motions in the atmosphere at low level and by the rivers of frigid air in rapid motion at higher altitudes. But if you can't see CAT, how can you describe it and how can you locate it?

Turbulence is familiar to anyone who has watched a lighted cigarette in an airless room. The bluish smoke will rise upward and as the heat of the smoke cools slightly it will undulate back and forth and gradually move outward in wavy horizontal lines. It may move up and down, or coil like a snake, dive and ascend and form thin layers, and yet there is not a breath to stir it.

This erratic behavior is turbulence. It is created by subtle changes in temperature, so insignificant that it is impossible to measure them. It is thought that minute temperature changes in the atmosphere likewise stir the air into motion and, because it is unseen, jets slamming into it at near the speed of sound get tossed around like corks. Sometimes they dive toward the earth and on occasion have actually broken the sound barrier before being recovered by their badly frightened crews.

There exists a criteria table to describe various degrees of turbulence. For instance, light turbulence is a condition during which aircraft occupants may be required to use seat belts but objects in the aircraft remain at rest. Airspeed fluctuations vary between five and fifteen knots and gust velocities vary between five and twenty feet per second from all directions.

Moderate turbulence is the condition in which occupants require seat belts and occasionally then are thrown against the seat belts. Unsecured objects in the cabin are thrown about, and the wind gusts can be expected to increase up to thirty-five feet per second. To most passengers this is uncomfortable and frightening. An unsecured passenger could be hurled against the ceiling of the aircraft, an occurrence that is not uncommon as nervous and sick passengers most often head for the toilets during this type of choppy air condition, and who can blame them? Flight attendants are forbidden to leave their own seats.

The official definition of severe turbulence is a condition in which "the aircraft momentarily may be out of control and occupants are thrown violently against the belts and back into the seats and objects are tossed about." Upward and downward wind gusts increase up to fifty feet a second but are not deemed severe enough to snap an airliner in two. But don't bet on it.

Extreme turbulence is lethal turbulence and although the Weather Bureau says that this condition is rarely encountered it has in fact happened on many, many occasions. Rapid fluctuations of airspeed are indicated, and up- and downdrafts are greater than fifty feet per second. Under this condition the "aircraft is violently tossed about and is practically impossible to control . . . and [the extreme turbulence] may cause structural damage."

These turbulence criteria were developed away back in 1957 before the U.S. jets were flying, but today it is a well-known fact that increased jet flying at all altitudes of our crowded skies leaves little room for circumnavigation of storms, and as a result severe and extreme turbulence is encountered more often than is believed. Sometimes luck alone decides the outcome.

Trying to solve a problem you can't see is indeed baffling, and a number of interesting ideas about turbulence have been offered. But first let's trace the CAT location around North America where the highest jet aircraft density is located. In the southwestern part of the United States where the land is arid and hot and where ragged mountains serrate the deep blue skies, a combination of heat and mountain currents seems to stir the lower atmosphere from stillness to life and start a river of air flowing northeastward from Arizona and New Mexico in the general direction of the eastern Great Lakes at a height of approximately 10,000 feet. This river of air is known as the lower jet stream.

Natives of New Mexico have often said that all the storms of the United States and Canada start in the vicinity of the Sandia Mountains near Albuquerque, and maybe they are partly right. Almost any spring or summer day, you can watch the clear skies over New Mexico gradually turn to clouds over Sandia Peak. By midafternoon, great boiling masses of black thunderclouds start northeastward across the Great Plains and head for their collisions with the humid air masses churning in from the tropics and the frigid air masses being swept southward from the Canadian Arctic.

With the exception of the storms of the Pacific Coast, which are caused by the impact of moisture-laden air against the mountains being swept up to great heights to condense and fall to the earth, the biggest and severest storms in the United States are caused by the combination of the central air movements just mentioned. A pencil trace along that line from Albuquerque to Detroit, Michigan, could best be described as "Tornado Alley," because along this line the rivers of air collide, sometimes softly, sometimes with great violence. The Kansas City Severe Storm Center, which predicts all the severe storms of the country, can pinpoint tornadoes and other severe storms by locating the presence of both jet streams when a low pressure area moves into a cold front associated with the intersection of these jet rivers. And along the right front quadrant of such an intersection, tornadoes will appear as well as other damaging winds with hail and lightning.

It was the big jets that located the high altitude jet stream, and pilots who once boasted they could speed up air travel by locating and riding the high winds of the upper air now sing a different tune. Jet streams are dangerous not just because they create Clear Air Turbulence, but because they contribute to other violence throughout much of the atmosphere.

The upper jet stream may be an abnormal situation according to scientists, but it's likely to be around for considerable time and the airlines will have to live with it. Originally in the first quarter of this century it undulated like a snake on both sides of the equator over the mid-temperate zones. But for some reason or other (some blame atomic bomb testing), the jet stream changed and in the Northern Hemisphere it swung far to the north until it struck the frigid Arctic regions where it turned southward over Alaska and the Northwest Territories of Canada, across the Canadian Plains, southward and eastward over the Kansas Plains to as far east as Michigan and south to the Gulf States where it was warmed slightly. It abruptly shifted northeastward over the Atlantic coast and went out over the ocean to just east of Newfoundland and thence to Iceland; here it abruptly turned again and headed south and slightly eastward over the British Isles, France, and Spain before being shunted northward again. And that is where it is located today.

This accounts for the severe storms of the Great Plains and the extreme Clear Air Turbulence experienced over the Central United States and off the Newfoundland Coast where air traffic is densest at great heights.

At the present time the upper jet stream can be haphazardly located by Weather Bureau radiosondes. As long as its boundaries are generally known and as long as the exact boundaries of the lower jet stream are available, the severe storms can be predicted with reasonable accuracy. There is only one hitch.

The upper altitude jet stream is undependable; it shifts back and forth. Perhaps it is the victim of solar flares, or perhaps it is attracted and moved by great belts of electricity over the North Pole, or given a shunt by the reflection of sunlight off the Arctic ice. In any event it follows an irregular path between 26,000 and 43,000 feet and perhaps even higher, as supersonic aircraft are expected to discover in the mid-1970s.

The industry now wants to find some means to detect the presence of the upper stream so that jet airliners can circumnavigate it, if possible. They must find out more about locating its presence, because at least one eminent authority on aerodynamics has predicted that Clear Air Turbulence will tear a supersonic airliner to shreds. So far the results of laboratory and high altitude experiments have been disappointing.

At first, a handful of scientists thought that laser beams of high intensity light could detect and report back on the presence of the jet stream, but laser beams could also melt an aircraft. This scheme was gradually given up, though it may return if some lock-in device can be found in the sky, a permanent fixture such as a star so that a beam trained on the star would report if there was any interference in its direct pathway. This may be a long way off, but still it has possibilities.

Eastern Airlines, which has no transatlantic routes where much of the jet turbulence is encountered, nevertheless has done considerable pioneering on the subject and discovered two years ago that turbulence occurred when air-temperature differences of as little as one or two degrees were encountered. Delicate instruments placed on an Eastern jet could sense these minute temperature changes taking place, to be followed in about two minutes by turbulence.

The interval between the start of the temperature change and the onset of the pitching was so rapid that a change in course was not possible. Yet, this was an important discovery, because heat-detecting instruments had already been perfected, particularly by the military who found the use of infrared scopes could detect subtle human body heat in the dense jungles at night. If infrared could detect such infinitesimal temperature changes on the ground, then why not in the air?

Tests by Pan American over a preliminary three hundred hour research phase in 1968 disclosed that infrared detectors could sense air temperature changes and Clear Air Turbulence up to forty and fifty miles ahead. Because of the amount of equipment required to make this detectional breakthrough, however, Pan Am safety engineers believe that operational equipment could not be available until 1971 or soon after. Satellite intercommunication systems to assist in locating and informing the teeming skyways of the discoveries of CAT will not be ready until the same time, and it is hoped both these systems will pave the way for greater safety in the coming supersonic era.

The jet calisthenics that awaken the industry to the dangers of Clear Air Turbulence occurred on a routine cross-country United Air Lines flight on July 12, 1963, which was followed by an Eastern Airlines incident on August 20 near Dulles International Airport and another Eastern experience on November 10 out of Houston, Texas.

The classic example was the United experience, not just because of a horrifying plunge toward the earth, but also because United circulated the warning of Clear Air Turbulence to all airline companies and to pilots' associations all over the world. Instead of hiding the experience, United wisely released a detailed study of the incident.

On that night of July 12, United Flight 746 swooshed off the runways of San Francisco International Airport at 6:25 P.M. bound for O'Hare Field, Chicago, and due there a few minutes after midnight. Aboard the Boeing 720B airliner, a sistership but slightly shorter than the Boeing 707, were fifty-three passengers and a crew of six. Far ahead of them, summer thunderstorms sailed like galleons across the evening skies, but between the clearly visible storms were vast areas of open sky and occasional cloudy patches in the deep distance.

It could be considered typical July flying weather, bumpy near the ground but cold and clear and relatively turbulence-free in the thin air to which the flight had been assigned,

37,000 feet. There is no greater feeling of exhilaration than flying a modern jet through clear evening skies surrounded by the sunset hues of distant clouds.

The U.S. Weather Bureau had predicted there would be a collision of cold and hot air masses that day along a line that stretched from Utah through Colorado and northeastward through Kansas and Nebraska, and onward on a line that ran just north of Chicago, attenuating somewhere in the northern part of Lower Michigan where the Great Lakes would dissipate the action.

Available to the pilots of this United Flight was a routine weather bulletin issued earlier in the afternoon, forecasting a familiar summer weather report: SCATTERED AFTERNOON AND EVENING THUNDERSHOWERS OVER THE EAST SLOPES OF COLORADO AND NEW MEXICO THROUGH EASTERN OKLAHOMA, WESTERN MISSOURI, WESTERN IOWA, SOUTHWESTERN MINNESOTA . . . WILL INCREASE DURING EVENING WITH THUNDERSTORMS BECOMING BROKEN IN SHORT LINES AND SPREADING EASTWARD TO CENTRAL WISCONSIN, EASTERN IOWA AND EASTERN MISSOURI BY EARLY MORNING.

As a result of this report, airborne radar as well as Air Traffic Control radar would reroute the jets flying the busy corridors between Chicago and the West Coast well away from the troublemakers.

Later in the afternoon the high altitude jet stream was discovered moving swiftly southward at an altitude of 37,000 feet, hugging the cloudtops that had been swirled around from dissipated thunderstorms.

Having located the jet stream, moving at upwards of 100 miles an hour, the Severe Storm Warning Center at Kansas City issued Severe Storm Warnings for an area in Nebraska surrounding the city of O'Neill; the warnings would be in effect until eight o'clock that night.

United's Flight 746 was at 37,000 feet, the level of the invisible jet stream, and its flight path was directly over O'Neill. Far below at the 5,000 foot level, a squall line was forming, typical of evening and nighttime thunderstorms in this region. As the warm air from the south was shoved aloft by the cold front, great masses of cumulus clouds thundered upward to 46,000 feet, some three thousand feet higher than the ceiling limits of a jet airliner.

United had always been a leader in impressing upon its flight crews the dangers of thunderstorms, and the crew flying the 720B that night were well aware of the danger and prepared to circumnavigate any storms on the flight path. The passengers were lucky to have such competent men at the controls. Captain Lynden Duescher, forty-two, was the pilot in command with an impressive 17,315 flying hours to his credit. First Officer Eric Anderson, thirty-four, had more than 10,000 hours behind him, while Second Officer Ervin Rochlits, forty-one, had about the same. Also on deck was another United Captain, E. P. Aiken, "deadheading" for Chicago. The group represented a century of flying and in a few minutes they would need all the experience they could muster to save their flight from disaster.

Near Scottsbluff, Nebraska, lightning was observed on the horizon to the south of the planned route. A storm echo about 130 miles ahead was located by the radar. As the flight continued eastward on automatic pilot, two more echoes were observed east of the first echo and about twenty-five miles apart, indicating the line of thunderstorms. When the aircraft reached a point fifty miles from the echoes, the radar was switched to its fifty-mile range for a closer scrutiny of the storm. Its "Cheerios" were fuzzy, indicating that the storm was mature, but the wind shears were not so sharp as could be expected from mature summer storms.

As the flight approached O'Neill, the nearest storm cell was due south and about forty miles away, but from the flight path a deck of cirro-stratus clouds was observed ahead with flat, thin tops which were familiar in form but never troublesome. Actually these thin clouds were scattered thunderstorm anvils that had been carried south by the high winds. The pilots noted that the cirrus deck sloped upward and skies were clear at the 41,000 feet level.

A slight chop was encountered as the jet entered the thin clouds. The captain asked Air Route Traffic for a clearance to 41,000 feet to get over the chop. There appeared to be no storms in the path and this was confirmed by three ground-based radars. The seat-belt signs were turned on in the passenger cabin. A slight increase in the airspeed of the jet was noted. First Officer Anderson disengaged the automatic pilot, took control of the wheel, and the airspeed was reduced to 250 knots, which is the rough-air-penetration speed for the 720B.

The crew was flying by the book. The anti-icers were turned on in case any ice should build up on the openings around the four jet engines suspended in pods from the wings. Light turbulence was now encountered and this increased to moderate turbulence and quickly to severe turbulence. The pilots had their shoulder straps on and both gripped the wheel. Clearance came from ATC to climb to 41,000 feet. But they would never reach it. They were caught in the Clear Air Turbulence of the jet stream which was aggravated by the clouds. The jet was now a roller coaster without tracks, threatening to get out of control.

The pilots attempted to climb the 720B but without success. A sharp downdraft caused a loss of altitude as well as a loss of airspeed. The nose was pointed upward into the cloud but the aircraft was dropping not climbing. The passengers by this time were badly frightened, and there was worry in the cockpit too. Forward pressure was applied to the control column by the pilots, and all four throttles were shoved forward to get more power in order to maintain airspeed and altitude. Jets stall quickly in turbulence at this great height. There exists only a slim margin between a high-speed stall and a low-speed stall in such thin air. The stall is caused by the knife-edge wings that are built for high speed to maintain their lift. Loss of speed meant loss of life and the engines roared a reply to the throttles.

The flight now entered a region of violent turbulence. The captain yelled that buffeting was occurring, the ominous shuddering that indicates a stall is beginning. As all fliers know, a stall occurs when an aircraft loses its air cushion of support and ceases to be a flying machine. As soon as the captain advised the copilot of the stall warning, Anderson abandoned his attempts to maintain altitude and concentrated on maintaining a level attitude; he manipulated his four throttles to keep his airspeed above the stall point and yet not high enough to create a structural problem in the continuing turbulence. He started an easy turn toward the northeast in an attempt to get out of the wind shears that were bouncing the jet like a rubber ball.

In his first attempt to gain height, Anderson managed to lift the jet only to about 37,800 feet, something he could have done at any other time in a few seconds. Now the aircraft had fallen to someplace around 36,500 and the second buffet occurred, occasioned by a tremendous updraft. The jet rose rapidly to 38,000 feet. The nose of the aircraft began lifting while the tail began to sink.

Both pilots shoved the control column as hard as they could to try and keep the nose down but to no avail. The nose kept going up. The jet was threatening to stand on its tail. Captain Duescher hurriedly informed the Denver Air Route Traffic Control that he could no longer hold altitude. The aircraft was out of control. The nose went upward to 40 degrees while all the while the turbulence was bouncing the jet up and down.

Loose articles began floating around the cockpit and in the passenger cabin as the trapped occupants began to experience weightlessness. The crew members were lifted from their seats but their shoulder harnesses prevented them from floating around the cabin. Books and papers flew through the air. Again the captain managed to call to Air Traffic Control that the aircraft was uncontrollable, leaving the controllers along the route with the helpless feeling of staring up into the night, unable to help.

At this second, the aircraft nosed over and dived. The instruments glowing in the dark told the story. The captain pulled on his speed brakes (flaps on the wings to retard the speed) as the speed of the diving jet reached over 400 knots, just ten points below the sound barrier. A warning bell was now ringing, indicating that design speed was exceeded, and it never stopped its unnerving message of doom.

The crewmen didn't know where they were at the moment, upside down, sideways, or what. The instruments were tumbling and the jet was spinning. Material was flying all over the cockpit. In the passenger cabin, pillows, blankets, magazines, and newspapers were vortexing over the aisles. And now another problem occurred. One of the passengers, a soldier, had failed to buckle his seat belt. As he flew out of the seat, passengers who remained strapped grabbed at his legs to hold him. Some reached for his jacket, others for his hair, and now began a struggle in a small eight-foot aisle area to save the careless traveler.

At 25,000 feet the hurtling jet, now believed to have broken through the sound barrier, slammed out of the thin cloud and both Duescher and Anderson were able to make out the lights of O'Neill far below. They knew now what the instruments couldn't tell them. They were diving earthward but the plane was not on its back. It was rightside up. The captain considered putting the wheels down to stop the hurtling giant but decided against it. They could be wrenched off. Then he thought of raising the air spoilers on the wings but decided that this might create a structural problem. He was thinking carefully and he didn't have much time left. Behind him his engineer was busy activating switches because of pressure oscillations in the cabin. Lights were flickering the danger signals on his control panel as he flipped circuit breakers, speeded up compressors, activated engine bleeding systems, and tried to balance the fuel pressure systems, while a myriad of warning lights flashed on his panel like a busy telephone switchboard.

The pilots began to ease back the control column. At one point a little power was boosted into the jet engines to help gain control by speeding up the air over the control surfaces, not an easy act to perform in a dive of this kind but necessary nevertheless. Down she dived . . . 20,000 feet . . . 16,000 feet.

The loose passenger struck an overhead rack and descended toward the floor in a prone position. Passengers hauled him into his seat, and he quickly found his belt and fastened it.

Up front, by pulling back slowly on the control column,

the pilots brought the nose up and jockeying the power plants brought the jet into a level position just under 14,000 feet. The captain called Air Traffic and advised that the jet was under normal control. He asked for a doctor to be at O'Hare to meet the arrival, but when the jet landed, only badly frightened people greeted the medical team. Investigation of the 720B showed no structural damage. The incident warned airlines and pilots around the world that Clear Air Turbulence could play strange tricks with jets. If the crew had not been so competent as it was, death could have been the ending to the story.

That was back in 1963, and Clear Air Turbulence problems have not been solved in the meantime. If the infrared experiments fail to work, what of the future as thousands more jets take to the skies? If trapped by CAT, they will be unable to move in any direction because of air congestion. It's not a pretty picture. Even over the vast Atlantic airspace is becoming seriously congested, while airlines and governments, opposed by the transatlantic pilots, try to narrow the distances between aircraft so as to crowd more flights into the lucrative skies, creating less and less room to circumnavigate trouble.

"When the weathermen studying the Atlantic charts tell us to expect CAT, we never run into it," says Captain James Foy, veteran pilot with Air Canada, past president of the Canadian Air Lines Pilots Association and one of the best-known safety experts in world aviation circles. He flies the Atlantic every day or so and says that no two flights are ever the same.

"But when they don't predict Clear Air Turbulence, that's when we get shaken to pieces," he added. "It just seems to be impossible to predict with any degree of accuracy, and the Atlantic airspace is now so crowded, you have a rough time trying to change course or altitude when you slam into such turbulence.

"For instance, on December 19, 1963, I was flying my DC-8 at the assigned altitude of 37,000 feet. The air was smooth and visibility was unlimited. I was eastbound, and the weather bureau at Newfoundland had predicted CAT at my altitude just east of Newfie. I was prepared for it; and my automatic pilot was off; my airspeed was lowered slightly.

"A Pan American jet was below me at 34,000 feet. There was another jet ahead of me and slightly lower at 35,000 feet. This will give you an idea of how crowded the North Atlantic is, these days.

"Suddenly, I heard the Pan Am jet reporting moderate to severe turbulence at his 34,000-foot level. It was all clear and calm where I was. It was clear with the other jet also. But Pan Am was right in the middle of the jet stream and began to report very severe turbulence, which is damn-well serious. The Pan Am pilot asked for a level change, but this was impossible because of the air traffic. So the Pan Am crew did the next best thing. They swung their ship on a ninety-degree turn to the south and, in that way, were able to eventually fly out of it.

"This is an example of how CAT works. You never know where it is. On my return flight, which would be on December 21, 1963, the Paris weather forecast was for jet streams over the North Atlantic from as low as 27,000 feet to as high as 47,000 feet, with moderate to severe turbulence just east of Newfie. We flew this route and found it smooth as a kitten all the way. It's like I said. When they predict it, you don't get it, and when they don't predict it, you run smack into it.

"And no wonder we get shaken! The upper jet stream swooshes along at something like 170 to more than 250 m.p.h., and the wind shear between still air and a stream of this kind is sharp indeed. We keep our fingers crossed," he said.

As another example of how Clear Air Turbulence affects the jets, take the case of American Airlines Flight 74, nonstop Los Angeles to Cleveland, Ohio, during a routine hop in January 1964. At the controls of the big 707-fanjet was Captain Herb Schmidt, veteran pilot for American. He was flying at 37,000 feet and was over the radio checkpoint of Farmington, New Mexico, close to the intersection of the state lines of Utah, Colorado, New Mexico, and Arizona.

Other pilots ahead of him suddenly reported Clear Air Turbulence between 35,000 and 41,000 feet. Captain Schmidt quickly called Ground Control and asked for permission to descend to 27,000 feet. He switched on the seat-belt signs in the passenger compartment, slowed his giant machine down to the recommended turbulent-air-penetration speed of 280 knots, grabbed the controls tightly, and waited for the onslaught, while Ground Control was trying to find him another altitude.

Suddenly, CAT struck like a load of bricks. It tossed the jet like a cork in an ocean storm for a full three minutes. Schmidt and his copilot fought to maintain their altitude and level attitude, while the jet pitched and bucked in what he described as "very severe turbulence."

Then there was a lull. A little old lady unbuckled her seat belt and rose up out of her seat and headed toward the washroom. The seat-belt lights were still on. Without warning, the lady was hurled up to the ceiling. As belted passengers leaned out into the aisle, holding onto her skirt and her legs, the jet bucked like a wild bronco. She hit the ceiling again and was badly injured before the passengers could get a grip on her bouncing body.

The flight broke through the turbulence several minutes later and went on to land in Cleveland. When Captain Schmidt went to turn off his Number Four engine (the power plant on the extreme right side), he could not get the throttle to move below idling speed, and he starved the engine by closing the fuel cock to make it stop. An examination of Number Four engine showed that its pylon had been twisted and had fouled the power controls.

The little old lady went to the hospital with a cracked vertebra but, by and large, she was lucky. Several other passengers were shaken but unhurt. Describing the experience after the landing at Cleveland, Captain Schmidt said that what surprised him and shocked him was the behavior of the instruments during the turbulence. Each instrument read differently. In other words, the pilot was reading his instruments differently from what the copilot was, not at all reassuring. The instruments' errors, of course, were caused by the violent turbulence associated with the jet stream, violence that could take a 1,500-ton giant and treat it like a yoyo.

How to fly such unseen turbulence without stalling or breaking the aircraft apart is a question that most pilots will argue about. Officially, the FAA and airframe manufacturers believe that it is better to penetrate Clear Air Turbulence, or any turbulence for that matter, with a reasonably high speed rather than a reduced speed. The reason for this strange agreement stems from the facts that Clear Air Turbulence has not yet been identified as the cause of a major air disaster and that jets over the Atlantic and over the Pacific have dived thousands of feet without structural breakup. So it seems better to slam into the turbulence at a higher speed than risk a stall.

There is only one hitch to this kind of thinking. The longer jets are in service, the more likely they are to bear the marks of metal fatigue, and a jet with metal-fatigue cracks cannot withstand the effects of severe turbulence as well as it could when it was newer. There is abundant evidence that metal fatigue has occurred in older jets and still is. In 1964, the FAA issued a "mandatory" notice to all airlines using the popular Boeing 707 and 720B jets to examine immediately the wing spars (the main structure support of both wings) to determine if metal cracking had occurred. Such cracks had been found in military versions of the same jets and military jets didn't get near the usage of the commercial ones. The Air Force had found much evidence of fatigue and had appropriated one and a half billion dollars to beef up their jet structures. Fatigue had caused the British Comet crashes and others and the FAA was concerned. At least the traveling public was spared the worry. Newspapers never mentioned the mandatory order.

5

Geostrophy and the Montreal Air Crash

There is another type of turbulence, lethal and ever-present, which, unlike Clear Air Turbulence, occurs near the ground level in any kind of weather. It is unseen, undetectable, and may have already been responsible or partially responsible for an appalling number of jet disasters, including those at Greater Cincinnati Airport, Montreal, and elsewhere. It may also have contributed to low-level midair collisions.

It is a disturbance that scientists would term mechanical convection or geostrophic convection, and it took the thin-winged jets to ferret it out. So little is known about this type of low-level turbulence that it has been estimated that not more than half a dozen weather experts are working on the subject. The warning of a University of Toronto scientist that governments should start investigation of this dangerous convection has so far been ignored.

Geostrophic convection usually happens at night and in rain showers and often in the areas to the rear of cold fronts. In simple terms it is wind that is traveling over the ground and is wafted upwards because of trees, buildings, escarpments, hills, and also from ravines that lead upward from deep river valleys. Winds of gale strength are hurled upward by these mechanical or geographical barriers and they become dangerous to jets during the takeoff and approach procedures when speeds are lower, turns are being made, and when no trouble is expected. Then suddenly the jets are into the ground, into a subdivision of homes, or they may even end up in a body of water close to the airport approach.

Airport planning could eliminate geostrophic convection problems, but mechanical convection (wind currents hoisted aloft by buildings) is difficult because as soon as an airport is constructed far out in the country, planning boards permit subdivisions and high-rise apartments to obstruct the flight paths. These buildings not only require new flying procedures, but they set up turbulence that is very dangerous to all those living above and below.

There is a possibility that the loss of an Air Canada DC-8F near Montreal on the windswept night of November 29, 1963, could have been caused by geostrophic convection and since no official cause was ever found to explain the crash, this seems one good educated guess. The disaster was one of the worst in air history, with the obliteration of 118 persons.

One of the reasons why no positive cause could be found was that at the time it was not mandatory for Canadian commercial aircraft to carry flight recorders. Therefore, this jet, unlike its sister jets in the United States, did not have a recorder installed, a situation that was quickly changed after the official inquiry into the crash. The flight also carried 51,000 pounds of JP4 fuel which contributed to the intense heat of the fire, helping to destroy the clues that had already been torn to pieces by the force of the impact. No bodies were recovered. But from the bits of human tissue that were scraped from the impact crater, it was determined that no fire or explosion occurred in the aircraft prior to its colliding with a Laurentian hillside.

Flight 831 was a regular "Friday-night businessman's milk-run" from Montreal to Toronto and points beyond. It was a popular flight because it was scheduled to leave Montreal at 6:10 P.M. It was a flight that was always filled to capacity, always with a waiting list, with stand-bys at the airport hoping to get seats from last-minute cancellations or from familiar "no-show" customers.

But Montreal and the district for many miles around was in

the grip of a bad storm that night, and ground transportation to the airport bogged down in the torrential rain showers which marked the forward progress of the storm. Some observers said it was the worst rainstorm in Montreal's history; worried TCA agents at the airport wondered whether the airport buses would make their run in time for the scheduled takeoff. The flight was delayed ten minutes in the hope that all the passengers would make it. But eight failed to get through the storm. Their limousine was snarled in traffic, a pileup so bad that it was taking cars twenty to thirty minutes to travel six city blocks. The flight could be held no longer, and agents filled the remaining seats with people who had patiently waited. The passengers entered the sleek silver aircraft through the front boarding door because deep pools of water surrounded the rear boarding-ramp area.

It was now 6:22 P.M. On the flight deck and in command of the aircraft was handsome John Snider, forty-seven-year-old veteran with TCA, a former RCAF bomber pilot, and a man with more than 17,000 flying hours to his credit. Captain Snider had flown commercially for almost eighteen years and was familiar with the DC-8 series, having flown 458 hours in them, as well as 103 hours in the new DC-8F that he would be taking to Toronto that night. In the copilot's seat was First Officer Harry Dyck, thirty-five, who had been with TCA since February 1953 and had chalked up 8,302 hours of commercial flying, of which 398 hours were in Douglas jets. The third pilot aboard was Second Officer Edward Baxter, twenty-nine, who had joined TCA in 1957 and had 3,603 flying hours to his credit, of which 277 hours had been accumulated in Douglas jets. Flight 831 had well-qualified pilots to shepherd it to Toronto—three men who knew the DC8-F perhaps better than any other pilots in the world.

Captain Snider, who lived in Toronto, had been on the reserve list that day, but the regular TCA flight from Düsseldorf, Germany, had been held up because of bad weather over the Atlantic, and this was the jet that normally would have been Flight 831 westward from Montreal. So Snider was asked to fly to Montreal and operate the replacement 831 back to

Toronto, at which point the regular crew from Toronto to Vancouver would take over and Snider could go home and join his family for the weekend.

On the flight deck, Snider and Dyck heard Air Traffic Control give them their route to Toronto. They were cleared to operate over the St. Eustache omnirange—a radio beacon for Montreal-to-Ottawa flights—and thence to Ottawa, southwestward to Kleinburg which was thirty miles north of Toronto and then south to Toronto Airport at a flight level of 29,000 feet, with instructions to report at 3,000 feet and at 7,000 feet on the climbout from Montreal Airport. Dyck acknowledged the clearance and brought to life the four mighty Pratt and Whitney JT3D turbofan engines, which could hurl 18,000 pounds of thrust each and lift an incredible weight of 315,000 pounds and fly it at 575 m.p.m. The DC-8F was known as the Cargo Trader because it was designed with huge loading doors so that it could be converted from passenger to cargo or carry mixed cargo and passengers, depending on the type of run and the time of the year. TCA found it good business to fly cargo in this jet when passenger loadings were down. TCA was the only airline in the world with the Cargo Trader, having just acquired the first four off the Douglas line. They had already lost one. It had been almost totally wrecked when its takeoff was aborted in London, England, with a full load of passengers gratefully spilled unharmed into a field as the aircraft caught fire.

The DC-8F pulling out of the Montreal loading area was 150 feet six inches long; 42 feet four inches in height, with a wingspan of 142 feet five inches, and a wing area of 2,868 square feet. Its fuel capacity was 156,760 pounds. Its maximum range with this fuel load was 7,090 miles. For this phase of the flight, the distance was only 315 air miles and would be completed in less than three quarters of an hour's flying time. Captain Snider heard the latest weather for his takeoff and climb-out: "Overcast with light rain and fog, visibility four miles, surface wind twelve m.p.h. from the northeast." The weatherman had earlier told the crew they could expect some light icing conditions and, as a result, the engine anti-icing system was turned on.

Reaching the south end of Runway Zero Six, Snider brought the four engines to full power, released the brakes, and began the takeoff. It was 6:28 P.M. In the passenger sections, purser James Zirnis and three stewardesses, Kay Creighton, Linda Slaught, and Lorna Jean Wallington, having checked the seat belts of all the passengers, buckled on their own and waited for the breathless thrill of the DC-8 takeoff. The aircraft streaked along the runway for almost two miles and then lifted its 213,000 pounds nicely into the rain-filled skies.

"Flight Eight Three One off at six twenty-nine," Dyck reported to the Departure Controller. His next report would be made when the jet passed through the 3,000-foot level. Montreal's northern suburbs sparkled through the light fog and rain. The speed of the aircraft would now be 175 knots, and below its shining wings and the bright lights of the cabin was the black water of Rivière-des-Prairies. At 6:30 P.M., the flight was just north of Laval-des-Rapides, whose lights could be seen as the aircraft entered the clouds and rain showers and disappeared from the Air Traffic Control radar. At 3,000 feet, copilot Dyck again reported, as he was supposed to do, and he acknowledged a lefthand turn permission to the St. Eustache omnirange.

The flight was now turning very slightly to the left, still climbing through heavy cloud and rain. If Captain Snider experienced any turbulence in his flight, he did not mention it either to the airport or to his company radio. The flight was expected to report going through 7,000 feet. In approximately four minutes, the aircraft was turning northwestward, crossing above Rivière-des-Milles-Iles and over the little village of Ste.-Thérèse in the foothills of the Laurentian Mountains. Four miles northwest of this village, something went wrong. The jet dived earthward and struck the ground with the shock of an earth tremor. In fact, the time of the crash was established by the shock recorded on the seismograph at the College Brébeuf in Montreal.

The flight was ended. It had traveled only 16.9 statute miles from the airport. Montreal had no knowledge of the crash, since the flight had been lost to its radar a few minutes before. It was the parish constable, Noël Aubertin in Ste.-Thérèse, who gave the alarm.

Constable Aubertin was in his patrol car when he heard the jet fly overhead in the rain clouds. He was driving slowly, he recalled, because of the rain and fog. He saw a boy and a girl standing by the roadway waiting for the provincial bus which would be along at any moment. He stopped the cruiser and advised the couple to stand farther back from the roadway as they were difficult to see in the bad weather. He had no sooner spoken to them than he heard the roar of the jet engines very close. Then there was a quick silence, followed by a tremendous explosion, and a flash of brilliant fire that lit the entire area. The night appeared to be filled with fire. Aubertin thought the world had come to an end. His first instinct was to shove the boy and girl to the ground. Diffused by fog and raindrops, the fire seemed everywhere. Regaining his senses, Aubertin put his thoughts in order, radioed the alarm to the Quebec Provincial Police through his cruiser radio, and asked them to notify the RCAF base at St. Hubert. Then he drove and later ran to the scene of the holocaust. He lost his heavy boots in the gumbo mud and ran on in his stocking feet, hoping to be of help.

The rain continued its furious tattoo but could not quell the flames that were shooting 500 feet into the sky, lighting up the Ste.-Thérèse area for 100 square miles. Aubertin could not get near the scene. The heat was just too much, and he stood helplessly by watching the great fire roar its requiem to the clouds above.

Trans-Canada procedures called for the DC-8 to take off the runways at speeds of almost precisely 170 knots, and climb at 3,000 feet per minute, reducing power on the climb to maintain 3,000 feet per minute, until reaching the altitude of 3,000 feet, when the recommended power setting would be increased to 250 knots at a climb of 2,500 feet per minute.

Ninety percent of all commercial airline pilots fly their jets manually during the takeoff, the climb to near cruise altitude, the descent, and the landing, leaving the use of the automatic pilot to cruising altitudes only. The others, who like to snap on the automatic pilot immediately after lifting off the runways, program the rate of ascent and the degree of turning into the control handle of the automatic pilot. Those who knew Captain Snider said he would be flying his jet manually on that night, while the other two pilots took care of power settings, communications, and establishing the position of the aircraft as it climbed and turned on its outbound course. In order to try to reconstruct the events of that tragic night, it was necessary for the investigators to know the behavior of the pilot and whether or not he flew to the "book." It was determined that Captain Snider flew by the rules, a man dedicated to the glowing instruments in front of him, one who would not allow his instincts to betray him.

At 3,000 feet, he adjusted his power settings, while Dyck called the Montreal controller. Beneath him, but hidden in the storm, was the town of Ste.-Rose. His indicated airspeed would be close to 250 knots, and his artificial horizon instrument would show that he was climbing steadily at about 2,500 feet per minute and turning slightly to the left as he activated his required turn to line up his jet with the radio course to Ottawa. There seems to be no doubt that the flight encountered some kind of turbulence in this area. It was certainly in the vicinity, and turbulence is always associated with the movement of a cold front through unstable air, sometimes severe and sometimes only moderate.

There was a possibility that a wind shear existed that night at the 6,000-foot level, and Snider would pass through this level just beyond Ste.-Rose. There was also the possibility that strong vertical air currents were present, due to the beginning of the Laurentian hills with their forested crowns. Wind currents, racing over the surface of the ground, were lofted upward by the hills and forests, and spaced between these currents would be the downdrafts associated with the heavy rain showers.

Since no definite cause was found, this next series of events is conjecture, but conjecture based on the behavior of other DC-8s during stormy flights. It is possible the jet entered an area of updraft, perhaps moderate to severe, caused by the twelve knot winds from the northeast striking the hills and turning upward. Pilots are reluctant to report turbulence unless it is sustained along their cruising flight path, when they will report its presence in order to get a change in their course that will take them away from it. Dyck made no mention of turbulence. Maybe there was none, but this is most unlikely. Maybe it was so severe that he did not have time to report it, too busy helping his pilot-in-command handle the controls and power settings.

In any event, Captain Snider, on entering the shear area would notice on his instruments, and would feel the sensation perhaps in his stomach, that the nose of the aircraft was lifting, and he would most likely activate his horizontal stabilizer to give the jet a nose-down trim to counteract the updraft. He would also call for a reduction of power for this penetration period.

On entering the downdraft, which would follow immediately after the updraft, the jet would instantly nose down, and because of its horizontal trimming to a slight nose-down position, the jet would now be placed in an accentuated nose-down attitude and would begin to dive and gather speed. The pilot's immediate reaction would be to give his aircraft a "nose-up" trim through the horizontal stabilizer, and then pull back on the control column to level off his flight. Unfortunately, jets begin to dive so rapidly that all the pressure in the world cannot pull back the column. The next reaction is to pull off the power, to slow down the hurtling giant. Next, if he has time, the pilot will activate his air spoilers on the wings to give the aircraft more drag. Then, if it is possible, he will even risk putting down his flaps to slow the aircraft, and then, if there are thousands of feet between his jet and the ground—say between 16,000 to 25,000 feet—he can put two of the engines into reverse and start the slow process of gaining control and pulling the jet into level flight.

But Captain Snider was only 6,000 feet above the ground,

and the jet would be hurtling toward it so fast, he would not have time to place all these "stoppers" in action. He would have time only to pull off his power. His instruments would be tumbling badly, and his ground reference, already lost in the cloudy night, would depend on instruments alone, and they would be going crazy. Therefore, he might not realize the degree of dive that his jet was taking, building up an incredible speed of some 400 to 450 knots. At speeds less than this, the pilots of a DC-8 over Texas managed to pull back the yoke with their combined strengths, and an engine was ripped loose. In this case, the speed may have been just too great, and the DC-8F, now with engines shut to idle, impacted into the sodden ground.

A cessation of power was heard by a ground witness, and the crash-investigation group found that the jet's power had been pulled off to idle-power setting at least ten seconds before the impact. This latter was determined by the throttle controls being frozen by impact at an idle stage, the setting immediately between the reverse power and the forward power. It was also calculated that the jet was flying about 470 knots—and maybe as high as 485 knots—prior to the impact, a speed very close to the sound barrier when computed in connection with the altitude at that area. Flaps and spoilers had not been activated.

The village constable, who heard the jet go overhead, was only four miles from the impact area. He heard the engines cease their roar, and almost immediately he heard and felt the impact. Stormy nights and heavy cloud formations often play tricks with human senses, but the evidence of this witness was important.

From the wreckage, it was found that the hydraulic stabilizer was set at an angle of between 1.65 and two degrees, nose-down trim, and had been operated to that position by the pilot with his horizontal stabilizer-control wheel, the very thing he would do if caught in an updraft, or if one of the plane's instruments gave him some misinformation. It was on this latter assumption that probers based their findings. Two pilots who were flying in this very area at the same time as the TCA jet, said that they were forced to slow down their aircraft because of severe turbulence. Despite their sworn testimony, turbulence was discounted as a factor by the commission of inquiry into the crash. The commission felt that the weather information, which was given in evidence, and the testimony of other pilots who flew in the area shortly before and after the accident, precluded the possibility of turbulence existing, "which in itself would be severe enough to cause the pilot any difficulty."

Therefore, if there was no turbulence severe enough to cause changes in the attitude (the best level position of an aircraft to slice the air stream as effortlessly as possible) of the jet that would require immediate remedial action, what caused the pilot to "down-trim" the nose? The only other explanation that would possibly stand up would be the malfunction of one of the stabilizing instruments in the cockpit, which would mislead the pilot and trick him into using a nose-down trim. But could a pilot as astute as Captain Snider be fooled by an instrument malfunction, when there was no other sensation associated with the necessity of requiring a nose-down trim—a sensation that can be felt by crew and passengers in any aircraft entering updraft or downdraft situations, situations often and erroneously called air pockets by passengers?

Maybe yes. Maybe no. But just in case it was yes, the commission of inquiry dealt with seven possibilities that could have happened and could have fooled the pilot-in-command into actuating the hydraulic stabilizer into that nose-down attitude. The seven possibilities were: failure of an airspeed indicator; icing or blockage of the static system (outside air at outside pressure to certain instruments, i.e., carburetor); leakage in the static system; unwitting engagement of the automatic pilot; failure or icing of the pitot system; erroneous indication of the aircraft's attitude; and the extension of a pitch-trim compensator.

It was the opinion of the experts at the inquiry that airspeed indicator failures were rare, and even if such a failure did occur, it could be detected by the crew before the jet would be placed into a dangerous, nose-down attitude. Icing or blockage of the static system: most unlikely. It would have been detected at the takeoff period. Leakage in the static system? Most unlikely. Automatic pilot engagement? A rare possibility, but the experts testified that it would be unreasonable to expect that the aircraft would be forced into an attitude and speed condition from which recovery was impossible. Failure or icing of the pitot system? The pitot is a hollow tube, sticking out from the leading edge of each wing, into which air is forced and calibrated to give the true airspeed of the aircraft. The pitot is heated so as to prevent icing, and if this heater should malfunction, there would indeed be a sudden drop in airspeed. If both pitots should ice up, then there would be no cross reference. But pilots and the experts at the inquiry thought it unlikely that both pitots would freeze up. The crew had been warned that icing was possible during the climb-out, and one of the first instruments they would activate would be the pitot heat, and they would refer to the instruments to make sure it was working. On this assumption, a failure of the system was deemed unlikely.

Having ruled out these reasons, the board next spent some considerable time discussing and evaluating the last two possibilities, the possible extension of a pitch-trim compensator and the erroneous indication of the jet's attitude. The pitch-trim compensator is a system that keeps a jet on an even keel when the aircraft is approaching very high subsonic speeds. A phenomenon takes place at these high speeds which causes the nose to point down slightly or, as pilots call it, "tuck under." This aerodynamic phenomenon forms shock waves over the aircraft's wings because of the high-speed airflow pattern, and these waves force the center of lift of the wings to shift slightly rearward, and the effect causes the jet to nose-down. An up-elevator movement is required to counteract the movement. The pitch-trim compensator is the instrument that moves the elevators to the correct angle as soon as it senses the critical airspeed where "tucking" can be expected.

This was done by a computer which triggers a twenty-eight-volt motor that moves a screw-type actuator on the elevators. If the pitch-trim compensator should malfunction and extend the full length of the screw mechanism, the pilot would sense this immediately by the pressure on his control column. If he was not aware of it, he might force the aircraft into an unwanted nose-down position, and there would be a tendency for the aircraft to pitch nose-down. Tests were conducted by the Federal Aviation Agency with a Douglas DC-8, and the results were submitted to the inquiry. They were to show that with a fully extended pitch-trim of nose-down, the pilot would have difficulty in maintaining the proper aircraft attitude, particularly in turbulence.

Despite the exhaustive tests on pitch-trim extension in conjunction with various horizontal stabilizer settings, leading United States pilots testified at another inquiry involving the death plunge of a DC-8 that a malfunctioning pitch-trim compensator would not cause any trouble. In fact, said one pilot, if the system was left out of the DC-8, the aircraft would do its job just as well and maybe better.

Charles A. Ruby, President of the Air Line Pilots Association, the organization to which 20,000 United States commercial pilots belong, testified a few months after the Trans-Canada crash that most pilots would rather have the pitch-trim compensators removed from their jets. Even if the compensators were improved, the pilots would still like to have them left off, Ruby said. One Douglas engineer said the DC-8 was safe no matter what the compensator did. Another research pilot then went on to state that a pitch trim was necessary, as it contributed to the stability of the aircraft, but he thought that jet accidents were more likely caused by turbulence or some other factor, and that accidents usually happened at night during climb-outs.

Even at this stage of jet flying procedures, pilots flying the same planes differed as to the importance of items such as the pitch-trim compensator.

As it appeared this was an unlikely culprit, this left only one other possibility, the erroneous indication to the flight crew of the aircraft's proper attitude. In the opinion of pilots reasonably well versed in the behavior of modern jets, this possibility could have been the one that sent the Trans-Canada jet on its death plunge.

The most important instrument before the pilot is the artificial horizon, a circular instrument with a floating ball dial which indicates whether the aircraft is in level flight, pointing upward, pointing downward, or banking to the left or to the right. It is actuated by a vertical gyroscope.

Pilots glue their eyes to the artificial horizon during takeoffs and landings, particularly in cloudy weather and in darkness, when all ground reference is lost. In airplanes big or small, in single engines or big jets, the automatic horizon is the most important instrument of all.

Some sophisticated instruments display the direction and deviation from the course, and some select the radio omnirange signals, the inbound localizer or radio-beacon course, the glide-slope indicator to the runway, and a warning if the gyro fails. The pilot knows that his gyro has failed by the appearance of a small red flag at the corner of the artificial-horizon dial. This warning was changed by Air Canada to a red warning light, because there was a gnawing fear that the gyro on the unlucky DC-8F might have failed, and either the failure was not indicated by the flag or the flag did appear and went unnoticed in the busy flight deck.

If the gyro failed and the red flag did not appear—and this is not an uncommon occurrence, according to pilots—the crew would be following an artificial horizon that could lead to the ground. It seems unbelievable at this stage of flying that any instrumentation of this kind is permitted. Yet it still is in many of the jets.

The red flag is also supposed to appear if the artificial horizon fails through a loss in electrical power, or if the horizon fails to follow the commands of the gyro. The worst situation occurs when the gyro fails and the artificial horizon follows the failed gyro, and the pilot, following the misinformation of the instrument, finds himself in a dive. At 6,000 feet, he would have little time to realize what had happened, and all the pulling in the world would not get the plunging jet back on an even keel at so low an altitude.

It is only through disasters of this kind that the public can know how inadequate instrumentation can be in modern eight-million-dollar jets upon which up to 180 lives or more must depend.

After exhaustive investigation, it was decided that the exact cause of the crash could not be found, but the most probable answer was the unprogrammed extension of the pitch-trim compensator. This conclusion was probably reinforced by the behavior of the Eastern Airlines DC-8 which plunged into Lake Pontchartrain. In fact, the Board of Inquiry into the TCA crash waited until the evidence had been taken at the Lake Pontchartrain inquiry before reaching a conclusion.

The Board of Inquiry recommended flight recorders on all turbine-powered aircraft in Canada. It suggested that DC-8 pilots should be made fully aware of the stability characteristics of their jets when the pitch-trim compensator is fully extended with the stabilizer trimmed to counteract this effect. An improved vertical gyro-warning system was recommended, as well as a more positive warning in case of a failure in the pitot heat system. The Board further thought there should be an improved means of indicating the horizontal stabilizer position to the pilots, and it suggested that a check list be called off by the pilots during takeoff and climbing procedures, as is done during landing procedures.

The Board included two wrist-slaps at TCA over maintenance procedures. It said:

It appears that the FAA Air-worthiness Directive (July 2, 1963) required that the elevator control-tab push-rod assembly be removed and visually inspected within 300 hours service time after April 18, 1963. Notwithstanding this directive, this inspection on aircraft CF-TJN was not made until 708 hours service time, and moreover the assembly was not removed but merely inspected in place.

It also appears that the FAA Air-worthiness Directive (January 24, 1961) requires that if any JT-3D3 engine was disassembled since last overhaul to the extent of exposing any bearing compartment, the main oil screen be inspected at periods of not more than twelve hours service time until the screen was free of contamination for two successive inspections. TCA inspected the main oil screens after ground running the engines and found them free of contamination but did not inspect them after time in service.

While the evidence does not indicate that either of these servicing shortcomings had any influence upon the crash, it is recommended that, in future, air-worthiness directives be followed and that appropriate procedures be instituted to ensure that this is done.

At Ste.-Thérèse, there had been too much confusion at the scene of the crash. Montreal television stations and radio stations were broadcasting minute-by-minute reports of the crash, some of them right, some of them wrong. The result of this confusion, and the failure of Trans-Canada to act quickly and positively to announce that the jet had crashed, drove thousands of curiosity seekers to the disaster site and left scores of crying and worried people in Toronto's International Airport confused and uncertain as to what had happened to their husbands and fathers. Not only that but hundreds of ghoulish spectators rushed into the mud and fire at Ste.-Thérèse to rob the belongings of the victims and steal away souvenirs of aircraft parts and structure that could have—and would have—assisted investigators in tracking down clues to the cause.

Ghoulish scavenging of aircraft sites was not confined to Quebec. It happens all too often at crash scenes. At Ste.-Thérèse, there was a clash between police and investigators as to who was in charge of the proceedings. Government and TCA officials, without police identification, were shoved around in the melee, while the evidence was disappearing. Then, to compound the confusion, heavy rains continued to fall, turning the site into a quagmire, which soon quenched the flames, buried the wreckage, and carried some of it away. Before the investigation could be concluded, it was necessary to build a giant cofferdam to protect the crews trying to find the wreckage, some of it sixty feet below the ground surface.

The disaster forced on TCA the use of flight recorders,

something that should have occurred at the same time the United States adopted them, in 1958.

Ste.-Thérèse acquainted investigators in the Department of Transport with the problems associated with jet disasters where the forces of collision and the heat of fuel fires would practically wipe out all clues.

This crash opened up an entirely new field in forensic medicine and the identification of human remains when only scraps of tissue and a few pieces of bone had been left in the ashes. Fingerprint identification was useless; dental charts, which were rushed to Ste.-Thérèse, were not needed. The force of the crash had exploded bone and enamel like a giant pile driver.

American scientists began working on the possibility of identifying humans by the molecular signatures of their skin and tissues. They explained that every object—organic, like humans; inorganic, like metals—gives off streams of molecules which can be identified and charted just like fingerprints. The trick was to find the infinitesimal difference in each signature so that identification could be positively linked to a particular person. Experiments for the FAA showed that entire bodies could be identified by the stream of molecules leaving them, but the tough problem was identifying bits of skin and microscopic pieces of tissue, although this work was continued with a fair amount of success. Molecular signatures were expected to be able to locate bombs in aircraft baggage by sniffing the air for the telltale molecules from nitrates and their compounds. For crash investigation, the sniffers would search the ground and the wreckage and then identify the objects, human and otherwise.

The investigation of the Ste.-Thérèse crash barely mentioned the geostrophic probability. It might have received more attention if the investigators had had before them the problems of geostrophic convection currents that were plaguing Greater Cincinnati Airport, and were soon to be prominently discussed after a series of disasters and near misses. The incidence of approach-to-airport disasters in Cincinnati spawned recriminations and lawsuits involving government agencies such as the Weather Bureau, the like of which had never occurred before. In fact, they are still going on. Pilots were no longer going to take the blame for their jets faltering on the final approach over such terrain conditions as found at the Cincinnati Airport, which incidentally is not in Ohio but in nearby Florence, Kentucky, a dozen miles south of the Ohio River.

On November 8, 1965, an American Airlines Boeing 727 plunged into a hillside while on its final approach to the airport and only four of the sixty-two persons aboard survived. Another crash occurred a year later on November 6 when a TWA 707 flight was aborted and one passenger died but thirty-six others escaped. Then on November 20, 1967, a TWA Convair 880 approaching the airport, again from the north and again at night and in the same month of the year, which may be significant, went down in an apple orchard short of the runway and, of the eighty-two persons aboard, sixty-nine were killed or burned to death while thirteen were tossed out of the wreck and lived.

Immediately after the last accident there was a move by pilots and by travelers who reside in Ohio to have the airport closed. It was no secret that landing or taking off over the north side of Cincinnati Airport was always an experience with stomach-turning turbulence as wind gusts whipped off the Ohio River, scurried through the tormented ravines south of the river, and then swooshed upward just short of the airport, which is built on a flat rise of ground, the only flat terrain for miles around.

So great was the feeling against this airport that the National Transportation Safety Board was forced to release an "Interim Report" on December 6, 1967, because of "the continuing flow of misinformation relating to this accident from sources outside the Safety Board . . . confusing to public and news media alike."

The Board then traced the flight of the TWA Convair 880 from Los Angeles International Airport to Cincinnati and disclosed that there was good visibility for the approach, with light to moderate winds from the east and some slight snow of no consequence. However, a runway extension of 900 feet to Runway 18 of the airport was not opened at the time of the accident and, further, the normal approach lights had been removed.

The jet struck trees about 10,000 feet short of the runway and bounced for some 3,000 feet before coming to rest and blowing up. There was no malfunction of the power plants nor the control systems, but the world's finest pilots just don't descend into the ground. The Board announced that it was concentrating on cockpit instrumentation and the systems associated with displays of altitude, rate of descent, airspeed, and pilot operational factors.

Imagine this type of investigation into cockpit instrumentation in this day and age, on the threshold of supersonic flying and rocketing to the moon! Yet the Board seemed to think the trouble was in the instrumentation despite the continued reports by pilots using Cincinnati that turbulence while approaching the runways from the north was always bad and could be dangerous at low-speed approaches.

While investigating this TWA crash the Safety Board was preparing the investigation results of the West Coast Airline crash of the new DC-9 on October 1, the year before. The sleek rear-engine jet with the T-tail configuration slammed into the Salmon Mountain while being radar vectored into the runway at Portland, Oregon. Thirteen passengers and five crew members were fatally injured and the aircraft was destroyed by impact and fire.

Most air travelers agree that the DC-9 is delightful to fly in, fast and quiet, with thrilling high-speed takeoffs and quick climbs into the smoother air of the high altitudes. On this night, the flight was assigned to 12,000 feet because it was only a short hop from Eugene to Portland. The jet crashed while under radar surveillance just eighteen minutes after takeoff.

Investigation of the cockpit voice recorder, the transmission tape at the control center in Portland, and the Flight Recorder of the airliner all showed that the crew knew where they were in regard to terrain and were flying perfectly in their approach to the airport. A number of instances of "sticky" altimeters were found during DC-9 simulated and "actual" approach flights, but examination of the altimeters in the burned airliner disclosed no malfunctions. Besides, the crew called out the altitudes and location of the flight and these tabulated exactly with the radar controller's observations.

The only significant item in the Flight Recorder read-out was an abrupt climb two seconds before impact. The Board determined that the probable cause was the descent of the jet below the surrounding mountainous terrain but the cause of the descent was not determined. It is interesting to note that the Board learned during the probe that the weather was generally described as low overcast with rain but "there was considerable variation in wind reported." "Moderate, occasionally severe turbulence just east of the Cascades" was forecast, and "elsewhere moderate turbulence within 4,000 feet of terrain where there were strong surface winds."

And the DC-9 crashed at the 3,830-foot level. Perhaps it's time the U.S. Weather Bureau started intense research in geostrophic convection. As pilots say: "Knowing is avoiding."

6

Flight Recorders

It has often been said that an airplane's flight recorder is to the investigator what the magnifying glass was to Sherlock Holmes. The flight recorder provides the clues to otherwise baffling airplane crashes. It keeps a permanent record of the behavior of an aircraft as it flies through the atmospheric conditions along its flight path. Its use is mandatory on all civil turbine-powered aircraft under the control of the United States Federal Aviation Administration. It is, likewise, mandatory in Great Britain and Canada and in certain other countries of the world.

In the days of the piston-engine airliner, investigators could find a host of clues in the pitch of the propeller blades, the position of the throttles, the trim mechanism, and the frozen records of the instruments. But thin-winged jet crashes occurred with such violence that clues were often destroyed, and only the presence of a flight recorder could determine details of the cause.

A flight recorder also helps the next of kin in airplane disaster deaths. Its story can be used as the basis for legal action against the airlines as well as against the FAA and the airframe manufacturers. It is no secret that if the flight recorder had not been made mandatory by government decree, it would never have been used at all. Airlines and pilots have never been amenable to the documentation of faulty flying practices nor to bad maintenance. When a flight recorder is totally destroyed by impact and fire, as happens all too often, the concern comes not from the airlines, manufacturers, or pilots but from the Safety Board or the FAA. The recorder is an unwelcome guest on many flights.

Though it has its flaws, the flight recorder has provided valuable information, and in so doing has made flying much safer. The case of the first BAC-111 is a good example of this.

The BAC-111 was an aircraft that promised profits and dependability for short, intercity hops. It was ninety-two feet long and twenty-three feet nine inches in height. It could move some 60,000 pounds at 500 m.p.h., at 25,000 feet, with its magnificent Rolls-Royce Spey engines, each engine developing 10,400 pounds of thrust. The two power plants were mounted on either side of the tail. The sweep of the wing was twenty degrees, not nearly as sharp as the bigger jets in commercial use. With fifteen degrees of flap, it could take off at 120 m.p.h.; and with thirty-five degrees of flap, it could settle down to a landing at 105 m.p.h. As a high-density carrier, it could transport sixty-nine passengers.

With a firm order for ten of the aircraft, the British Aircraft Corporation went into high gear in May 1961, and the first aircraft was rolled out in the spring of 1963, at Hurn, in Southern England. Then, after many months spent on in-flight evaluation, the aircraft met with disaster. On October 22, 1963, it literally fell out of the sky, taking to their deaths some of the finest fliers and aircraft engineers in England. It was a blow of the highest magnitude to the British aircraft industry. There were not a few smirks on the American side of the Atlantic, although Douglas Aircraft was very much concerned because the coming DC-9 was similar in many respects.

Because of its flight recorder, the British Air Ministry was able to learn the cause of the crash within two days. It was the fastest accident evaluation in aircraft history. The report showed:

> Information extracted from the two flight recorders recovered after the accident has now been analyzed in detail and shows that, during this test, the incidence increased substantially above the figure anticipated. The flight recorders show that the G-break at

the stall was large and abrupt, causing downward acceleration and further rapid increase of incidence.

The angle of incidence is the angle between the chord line of the wing (running through the leading and the trailing edges) and the direction of the flow of air. It is often called the angle of attack. It means simply the angle at which the wing meets the air through which it is flying. The G-break is an aerodynamic condition during a stall, when the lift can no longer be maintained equal to the weight. In other words, the plane ceases to fly. Downward acceleration means the rapid sinking of the airliner with low forward speed, which causes the airflow to be coming from below the plane rather than from the front, as is the case in normal operation. This underneath flow greatly increases that angle of incidence. The nose goes up.

A condition rapidly developed in which it would be impossible for a pilot, even of Lithgow's calibre, to appreciate the situation soon enough, and therefore prevent the further build-up of incidence.

As the incidence increased due to downward acceleration, the elevator started to trail up. This trail up was arrested and partial down elevator applied some three seconds after the G-break. The aircraft response, as could be expected under these conditions of low forward speed and aft CG, was too slow to stop the further increase of the incidence.

Eventually [in seconds] the incidence increased to such an extent that the effectiveness of the tail plane and elevator was reduced to a fraction of its normal value. The aircraft continued to descend at a high rate of descent, the fuselage attitude being substantially horizontal, and hit the ground flat.

The aircraft did not spin. There was no evidence of structural or mechanical failure. The flight-recorder data indicates that the engines were running, and were used during the attempt to recover from the stall. The wreckage indicates that the engines were still revolving when the aircraft struck the ground, and no evidence has been found that would indicate any in-flight malfunction.

This concluded the report of the Chief Inspector of Accidents for the British Ministry of Aviation. It is reprinted here because its flight recorders told the whole story within hours after the tragedy. The report disclosed what happened when a modern jet airliner stalls in flight under the strange circumstances that are peculiar to swept-wing aircraft.

As a result of the crash, the British took steps to remedy the situation. Alterations were made to the leading edges of the wings to improve the nose-down pitching characteristics of the aircraft. Modifications were made in the elevator linkage to allow a more direct connection between the pilot's control column and the elevator, thereby permitting the pilots in the future to override the tendency of the elevator to trail upward.

The costs of such modifications would be high. And the delivery dates to British, United, Braniff, and American would be delayed. But all this was well worth the time and the cost because the BAC-111 would be a much safer aircraft.

Flight-recording apparatus is manufactured in the United States and England. Three American companies prefer the use of aluminum foil on which to record the data, while two others prefer stainless steel. The recorders most commonly used at the moment are those with narrow aluminum tapes that are about the same thickness as household foil. One recording manufacturer is using magnetic tape, and this may someday replace the steel and aluminum.

It has been argued by some airlines, and by pilots, that present-day recorders leave much to be desired and do not tell the entire story, leaving wide gaps for conjecture. Present American recorders measure five parameters only—the aircraft heading, altitude, time, acceleration, and airspeed. A British-made system has a potential measuring system of some 200 parameters.

Flight recorders are not expensive, slightly more than $5,000 each. The tape—whether aluminum or steel—is reeled mechanically from one cylinder to another, not unlike the film in a movie camera. As it travels from one reel to the other, it passes beneath a number of steel styli (or needles) which indent the foil with a fine permanent record. There is a stylus for each parameter but not for the time sequence, which is marked by notches that show seconds and minutes from takeoff to landing. The styli move up and down over the foil tracing almost imperceptible scratches of the aircraft's behavior, its twists and turns and ups and downs. The needles are placed in such a way as not to strike one another as they are guided by remote instruments over the face of the foil. The recorder investigation of the Pan Am disaster at Elkton is ideal to explain the read-out of a flight recorder. When the wreckage of the jet had cooled sufficiently, the morning after the crash, a CAB team began looking for the recorder in the tangle of twisted metal. On the first intensive go-around, it could not be found. But the next sweep of the junk pile that was scattered over the muddy field filtered out every scrap of metal, bit by bit. When the recorder was finally located, it was almost unrecognizable. Only the telltale coloring of the heat-resistant paint on its broken shell revealed its identity. No longer a sphere, it was a blackened, crushed glob of metal. When it had been built by Lockheed, it had weighed thirty-two pounds and was fifteen inches high. To the investigators it looked like a walnut that had been beaten by a hammer.

The experts were extremely careful in lifting the crushed recorder. As is often the case after a crash, the styli are imbedded in the tape, and any violent movement might cause them to scratch further in the tape or tear the foil, thereby changing the reading of events. Just as carefully, the battered instrument was hand-carried to Washington to the CAB headquarters where it was dissected by Orion "Ed" Patton and his fellow engineers in the Bureau of Safety.

With infinite patience, they chipped away at the heavy steel outer shell, probed through the clay heat barrier, and scraped into the inner steel shell where the recorder lay flattened like an eggshell. The tape was there, but it was shredded and chopped into thousands of bits and pieces, an indication of the tremendous force of the crash and the vulnerability of the recorder's strength and location. It was immediately obvious to Patton that an entirely new course of attack was necessary if anything was to be learned from the frightful collection of aluminum chips. No flight recorder tape had ever looked as bad as this one.

But Patton was an old hand at tape read-outs and so were his staff members. They knew that each stylus makes its mark on the same piece of aluminum but that each scratch is microscopically distinct from the marks made by other styli. In other words, like fingerprints, each stylus had its own identification in the strange world of microenlargement. The trick would be to identify the mark of each stylus like ballistic scratches of a bullet jacket and then examine each scrap of foil and lay it out along its own peculiar stylus line. It would be a microscopic jigsaw puzzle.

For instance, the altitude trace would be identified. Then all the aluminum bits that had this trace would be assigned to the altitude path. The same procedure was then applied to the attitude, heading, and acceleration traces. Starting at the time of the aircraft's takeoff, each bit of foil was carefully set in place on adhesive tape, from left to right. It was an enormous task. Under the microscope, the aluminum took on the appearance of a window pane that had been newly etched by frost. It looked like a silvery panorama of roads and trees, of flowers and ferns. Through it all ran the tiny needle traces of the styli.

Patton and his men learned to match fern with fern, branch to branch, and stylus lines to stylus lines—with bits and pieces that could not be seen by the naked eye. They started their work late on the night of December 9, and by working eighteen hours each day, managed to complete the jigsaw by December 21.

When it was finished, the tape measured 4.2 inches in length, and it covered the last thirty-five minutes of the Pan Am flight. As it had been assembled microscopically, so it had to be translated. The CAB had designed an optical system with four binocular lenses which could report to a viewing screen all measurements up to one ten-thousandth of an inch. It had cost $15,000 to build. The findings on the screen would be mathematically translated to a large scale graph, as long as five feet in length and as wide as thirty inches. Meanwhile, computers would assist in analyzing the tracings.

Thus, the story began to unfold from the infinitesimal

scratchings. Measurements were made along both horizontal and vertical axes and were transferred to digital counters to get their relationship to time, altitudes, attitudes, and so on. As the information was correlated, the read-out showed that the 707 had traveled for thirty-five minutes and that it entered the holding pattern at 5,000 feet, turning, as it was supposed to turn, in three circles at a speed of 340 knots. There was no severe turbulence whatsoever, and there was no need for the pilots to slow down to recommended turbulent-speed penetration.

A small disturbance was noted at one place on the altitude line, and then the altitude trace dropped sharply to the bottom of the tape, which in reality was the ground. Something had caused an interruption in the level flight of the airliner, and a split second later its level changed and it dived earthward. The interruption was the lightning strike with an instantaneous explosion. It was all there, clear to the CAB sleuths. Had there been no such tape, no recorder aboard the doomed craft, it might have been difficult indeed to arrive at such a conclusion.

But the tape was a permanent record from a system that had been hotly opposed by the airlines from the beginning. The CAB admittedly would have liked to have had a system patterned after the British, with many more parameters, but the five parameters in current use were better than nothing. The CAB was grateful that the airlines had accepted the use of recorders, recalling that when in 1940 they attempted to install recorders into DC-3s, they were curtly dismissed. They tried again, when the Constellations were introduced, and later with the DC-6s. The tremendous size and weights of these airliners, and the prodigious amounts of fuel they carried, made it difficult to find clues in their wreckage. But the airlines were against any system of "spying."

However, in 1958, the CAB broke the stalemate and the FAA issued a "Technical Standards Order, Number V-51" recommending that a flight-recorder system be implemented in all jets. There was an immediate howl by the airlines because it meant added cost among other things. But the order held, and to assist the airlines in arriving at a cost-location decision, the CAB permitted a choice of three different flight recorders and of three possible locations within the airframe of the jets. After a decision as to which recorder to use, each airline then had to choose where to locate the instrument. It could be: (a) anywhere in the aircraft which would not upset the structural configuration of the jet and would not require special maintenance procedures; (b) an installation limited to a particular section of the aircraft which would be highly protected from fire and destruction, such as the tailpiece section; (c) an installation on the aircraft in such a place and manner as to be ejected seconds before the airliner impacted with ground, sea, or mountaintop. This third type was to be colored bright orange and would emit radio signals.

The airlines settled on the first choice, and the flight recorders were purchased from various manufacturers and placed in the wheel well where servicing was much simpler than at the top of a tailpiece. The fact that this location placed the recorders directly under massive fuel tanks that would subject them to long periods of intense fire did not seem to matter. The CAB felt that half a loaf was better than none and sat back content for the moment.

Unfortunately, there had to be a rash of other crashes with losses of their flight recorders before the CAB made its final move. A Pan American 707 was on its approach to the airport on the island of Antigua in the West Indies during a wild tropical storm on the night of September 17, 1965, when it collided with Chances Mountain at the 2,600-foot level, and all thirty persons aboard died. The flight recorder was destroyed completely in the fire that followed. The flight recorder from an Eastern DC-8 that plunged into Lake Pontchartrain, Louisiana, on February 18, 1964, was so badly mangled it was unreadable. The flight recorder on a United 727 jet that plunged into Lake Michigan on August 16, 1965, could not be found, although most of the wreckage was located and hauled to the surface. The flight recorder on an American Airlines 727 jet that crashed near Cincinnati on November 8, 1965, was destroyed by the intense heat of the fire.

These crashes and others proved that the flight recorders

were not sufficiently strong to escape destruction and were located in areas where intense fires were likely to erupt. The CAB, early in September 1965, recommended that the recorders be relocated just forward of the rearmost cabin pressure bulkhead. Nothing resulted from this advice, but other jet crashes forced the CAB to prod the FAA, which is the regulatory agency, to make a move. Finally, in January 1966, all airlines and other commercial operators were ordered to install or relocate by December 15, 1967, all flight recorders as far back in the aircraft as practicable for maximum protection.

Included in the new FAA rules was a revised Technical Standard Order which required a flight recorder to survive a thirty-six-hour immersion in sea water, 1,000 Gs of impact force, and to prove its resistance to penetration equal to dropping a 500-pound steel bar on the recorder from ten feet. Static tests under the revised order called for resistance to a 5,000-pound crushing force applied to each of a flight recorder's critical sides for five minutes in sequence. Fire-resistance standards would now require a recorder to survive 1,100 degrees Centigrade in which at least half of the outside area of the case was exposed to flame for as long as thirty minutes, depending on the type of recorder. Henceforth, recorders would be colored bright orange or bright yellow, so that they could be found more easily on the ground and in the water.

Then, to add more teeth to the order, the airlines were told they would have to conduct accuracy tests to show the correlation between the recorder data and the readings on the pilot's instruments. These strict rules were long overdue. It took jet crashes to make them mandatory. The next rule soon to be made will make recorders self-surviving. Before impact they will be exploded from the aircraft and parachuted to the ground. Perhaps someday the passengers will get the same consideration.

7

Survivability

The big jets introduced a new word into accident nomenclature. It is called survivability. It refers to an accident, usually on an airport, where the passengers survive the incident and should be able to escape the fire that inevitably follows an aborted takeoff, a bad landing, or a catastrophic engine or structural failure. This type of disaster is referred to by the Safety Board as a survivable accident but since, almost without exception, passengers have been trapped and have died in survivable accidents the word has taken on a grim meaning. Those who should have lived have not. They have died because jets cannot be evacuated quickly enough due to outmoded safety standards and inadequate fire protection, both in the aircraft and on the airport.

A survivable accident is a horrifying accident because it means that the victims have either burned to death or mercifully asphyxiated before the flames have reached them. It is all the more shocking because a survivable accident should be and can be survivable. It is one of the main goals of the Department of Safety and the FAA to make aircraft and airports safer, but no matter how they fight for the cause they are overwhelmed by the airlines and the industry, who do not want government interference in how an airline should be operated in the air and on the ground.

It is a struggle which can only end in favor of the passengers. A survivable crash that claims even one life on an entire airbus loaded with hundreds of lives will not and cannot be tolerated: the government agencies are just as adamant as the airlines and the manufacturers.

Sometimes it doesn't seem to be a question of what's right. It appears to be a question of economics, and airlines are willing to suffer "acceptable losses" as well as millions of dollars in claims payments to maintain absolute command of their air fleets and the dollars that every passenger mile brings over each hour of jet utilization. Removing twelve seats in a Boeing 727, to make aisle accesses to four over-the-wing exits available to passengers in a hurry, means fantastic losses in revenue. To remove some forty-eight or more seats in a jumbojet to permit freer escape routes will require an act of Congress with teeth to enforce it.

In the very face of a series of government proposals to improve the escape mechanism of a standard jet, manufacturers and airlines placed stretched-jets in service, added another fifty to seventy passenger seats without increasing the number of safety escape windows and doors. This contemptuous slap in the face to the Safety Board and the FAA could have been accomplished only by the tremendous political impact of the industry and the economic threat that is always dangled, with appropriate floodlighting, when anything or anyone interferes with the way that airline business is operated.

Yet, without tax dollars, the airlines could never have gotten airborne. Many still cannot operate without public subsidy, and without public monies the supersonic program will expire. It's time both sides got together. The solving of the safety gap by making survivable accidents survivable will be an important step.

The government agencies recognize that safety has a time limit, that a jet must be evacuated in ninety seconds or less if survivability is to become a reality. Forty-five seconds would be a more reliable figure but the government doesn't want to move that fast. However, if ninety seconds is deemed to be the time allotted between life and death, it should mean that airport fire and rescue facilities must also be in action in less than ninety seconds—and this, under present circumstances at all U.S. airports, is impossible, even if a disaster-marked jet stops at the front door of the fire station.

In the first decade of the jets, U.S. flaglines have been involved in seventy-one fatal jet accidents in which 954 persons have died. Seventeen percent of these accidents could have been survivable and 25 percent of the casualties might have been saved. In 1961, eighteen passengers of a DC-8, which wheeled off the runway and collided with a truck on Denver's Stapleton Airport, died because they were overcome with smoke. At Salt Lake City in 1965, forty-three passengers of a 727 unharmed by a hard landing, died because they couldn't escape in time to beat the mounting fire. And as recently as 1968, U.S.-built 707s have been involved in survivable accidents at Calcutta and London and passengers have died in the flames, in the London case with fire apparatus at the side of the aircraft.

Obviously, safety on the ground is as important as safety in the air. On the night of November 11, 1965, Flight 227 of United Air Lines, a milk-run from New York City to San Francisco, was speeding above cloud-filled skies at 31,000 feet, on its fifty-seven-minute leg from Denver to Salt Lake City.

Flying the aircraft was First Officer Philip E. Spicer, thirty-nine, a copilot with over 6,000 flying hours to his credit but only 84 hours in the Boeing 727 that he was now shepherding through the gathering darkness. At his left was his skipper, Captain Gale C. Kehmeier, forty-seven, a pilot with United since 1941 with an impressive 17,743 hours of pilot time including 334 hours in the controversial 727 jet. Second Officer Ron Christensen, twenty-eight, was also on duty on the flight deck. He had 166 hours in the 727, an aircraft with three rear-mounted engines that was in the process at that time of making a bad name for itself.

Back in the huge passenger cabin of the jet, eighty-five passengers were being taken care of by the three stewardesses. In a few minutes, only forty-eight of the ninety-one persons aboard the aircraft would be alive, the rest would be the victims of a fire-drenched accident which they should have survived. Two of the survivors would die later.

Everything about this flight was normal and routine. It was 7:35 P.M. and at the moment the flight was being advised by the Salt Lake City Air Traffic Control Center that "at the pilot's discretion" the flight could descend to the Salt Lake City

Airport whose 10,000-foot runway, 34-Left, was lighted and awaiting the arrival of the jet.

Captain Kehmeier called Air Traffic: "Let me know when we're sixty miles east of Lehi"—a radio intersection twenty-three miles southeast of the airport. At 7:38 the Air Traffic Controller informed the flight that it had reached this mark and the captain replied: "Okay, we'll start her down." The first officer now instituted a rapid descent toward the radio markers that would guide the jet to the threshold of the airport.

Four days previous to this, a revised routing procedure had been instituted by the airport bringing flights approaching from the east over the town of Lehi as an alternative to flying much further south to the town of Provo. By coming over Lehi, the approach to the airport was shorter by several miles and the descent was therefore steeper.

The beautiful swept-wing jet dropped through the skies at 370 knots and penetrated some heavy overcast about 6,000 feet in depth. The engine anti-icers were turned on, the throttles retracted to idle thrust and the speed brakes were selected. At 11,000 feet the speed brakes were retracted and there was a visual reference to the airport through the scattered clouds ahead and below. The flight continued down as 2,000 feet per minute and at 10,200 feet, the descent was slowed down as the aircraft was only four and a half minutes from touchdown.

"Okay, we've slowed down to only two fifty [knots] and we're at ten [10,000 feet], we have the runway in sight now," radioed the captain. Control of the flight was now switched to the tower and at 7:49 the landing instructions were given.

At 7,800 feet the 727 jet reached its stabilized approach speed of 123 knots, but it was dropping at a tremendous downward rate known as a rapid rate of sink, reaching an excess of 2,300 feet per minute when the United Air Lines' recommended approach speed at this moment was between 600 and 800 feet per minute.

Clearly the flight was sinking too rapidly. The first officer started to apply power when the captain brushed his hand away and said "not yet." They were now at 6,500 feet. Twenty seconds later at 5,500 feet the throttles were moved forward but the engines did not appear to respond and the captain immediately assumed control, moving the thrust to full takeoff power.

The jet was now only one and a quarter miles from the runway at an altitude of 1,000 feet, ready for the flareout and the landing. There were only thirty seconds left in the flight. The engines were now responding to the full thrust of the throttles but it was too late. Some 335 feet short of the threshold of the active runway, the jet struck the ground and immediately the main landing gear began to separate under the tremendous shock of 4.7 G-forces. Passing the threshold light, the aircraft slammed into the runway and continued to slide on the fuselage and the nose gear for 2,838 feet veering to the right for 150 feet off the runway while the Number One engine of the jet kept right on bounding over the field for another 140 feet.

At precisely 7:52:12 the tower controller yelled, "United's on fire . . . just landed," and the alarm was sounded.

As usual in such accidents, the passenger cabin was all confusion. And no wonder. Ruptured fuel lines in the belly were spewing flames all around the aircraft. Smoke filled the cabin and the emergency lights, which were supposed to illuminate the emergency exits, failed. All the regular cabin lights went out and passengers began shoving for the pitifully few exits on the 727.

Orson E. Nebeker was a passenger in the rear section, the section reserved for tourists, as it is in all jets. A Salt Lake City fire department dispatcher, he had been on a visit to Denver and this was his first jet flight. "Like a tourist" he noted everything he could about the interior, and "like a fireman" he noted the position of the two window exits over the wings. In his seat booklet, he had noticed that there was an exit just forward of his cabin bulkhead and referred to as a galley doorway. Noting these things saved his life.

Nebeker said that fire broke out immediately after the wheels collapsed, and a terrific blast of heat rushed into the passenger compartment. Realizing that the dangerous heat fumes would envelop him in a second, he took a deep breath and held it, then groped his way forward past screaming passengers, some of who were trying to escape through passenger-jammed window exits and some through the rear door. The lights remained on in the aft portion for a few seconds enabling him to find his way up the aisle. Nebeker groped ahead holding his breath as people dropped beside him, clutching at their throats. He found the bulkhead with his outstretched hands, reached around and located the galley door, opened it and escaped. The entire aircraft was now enveloped in flames and he began assisting other people to escape, as superheated gases cut down many possible survivors.

In the forward section of the jet, stewardess Virginia Cole rushed to the main front entrance, but she was crushed against the door by a group of shouting, pushing men. First Officer Spicer came out from the flight deck and shoved several of the men back, and then the stewardess was able to swing the lever and open the door. She was carried forward in the rush. "I tried to get back but they restrained me, yelling that I would be 'a fool' to try and help the others," she testified before the Civil Aeronautics Board public inquiry. One passenger by the name of Arnold escaped from the aircraft and then went back inside the main front door to help the trapped others. He found an unconscious man lying against the cockpit door, blocking Captain Kehmeier's attempt to escape from the flight deck. Arnold carried this passenger outside.

Meanwhile, Stewardess Annette Foltz, who had been with the airline just over a year, unfastened her seat belt in the rear jump seat and with two male passengers who were seated in the last row, attempted to escape through the rear doorway, which in the 727 descends to the ground under the tailpiece. Although this ventral doorway is listed as an escape exit and is listed in the Standard Operating Procedures Manual of *Aircraft Rescue and Firefighting* as an emergency doorway, it is anything but. It is impossible to use in a belly landing or any wheels-up crash, as almost all crashes are. But two men and the stewardess didn't know this as they activated the stair-descent system. The attitude of the aircraft prevented its opening more than six inches.

The three then attempted to back up but were trapped by the flames in the passenger cabin. They huddled close to the floor in the stairwell and as far back as possible from the approaching fire. Outside, the airport fire department had arrived.

Captain David A. Barrett was in charge of the equipment and he estimated the run at three minutes. The fire was so extensive that the thought was running through his mind: "Nobody will get out of this one." He said he entered the burning plane once and then donned a mask and tried to re-enter with his foam hose but gave up because of the intense heat. He said the nose-up attitude of the plane acted like a chimney and the hot gases exploded into flame with great rapidity.

Then the airport firemen ran out of water and had to await the arrival of the Salt Lake City fire apparatus that reached the scene ten minutes after the crash.

As the intense fire worked rearward, it burned a hole in the fuselage, near where the stewardess and the two passengers were huddled while firemen outside believed that everyone was dead inside. It was the waving of the stewardess' small hand through this tiny hole in the fuselage that attracted a passenger's attention, who in turn informed firemen there were people alive in the aircraft. Firemen shoved a small inch and a half hose nozzle through the crack and one of the trapped passengers hosed down the rear stairwell, all the while huddling on the floor.

Then came the dilemma . . . how to rescue those trapped in the tail. At the side of the aircraft was fire equipment considered to be equal to the best fire equipment on duty at U.S. airports and supported by a modern city department. However, it was ill-equipped to fight a modern jet fire and effect rescues from the furnacelike interiors. It was bad enough running out of water and having to wait until more arrived from Salt Lake City.

It took an incredible twenty-three minutes to discover there were survivors and then, lacking equipment to rescue them quickly, it took an unbelievable twenty-five to thirty minutes after the accident to effect a rescue.

Why?

Because the combined ax-wielding firemen of the airport and city departments didn't have axes big enough and heavy enough to chop through the heavy ribbing and the tough duraluminum of the aircraft. Repeated assaults upon the surface only caused the axes to bounce off, and it was impossible to slash into the thick ribbing of the 727 tailpiece.

As a result, they had to wait for the fire to burn a hole big enough to get the three survivors out. No wonder one of them died two weeks later from burns. The Civil Aeronautics Board, never a group to use adjectives, described the activity as "unprecedented."

The CAB investigation disclosed that this was a "survivable accident," but forty-three people did not survive because of a combination of dismaying circumstances that have not appreciably changed since the Salt Lake City disaster and are not likely to change until an entirely new approach is taken to suppress jet fires and effect passenger rescues.

At an FAA hearing in Washington in March 1966, one of the survivors of the Salt Lake crash testified before a gathering of delegates from airlines and industry who were hoping to find ways and means to improve survival rates. R. H. Dawson, a forty-four-year-old explosives engineer from Wilmington, Delaware, said that passengers had opened five of the six exits in the 727, but fumbled with three of them before they could get them open.

"No one will ever know how many lives were lost because of these few collective seconds' delay," he told the group. He said his business associate, a former champion athlete, ran by five of the exits to be trapped by the smoke in the forward part of the plane, and died there with twenty-nine others in the front.

He attributed his own survival to the fact that at Denver he looked hard at the nearest exit windows and when the crash occurred he focused his eyes in the direction of the window and headed for it through the smoke. He offered a suggestion. He thought that emergency areas adjacent to escape windows should have roughened surfaces on the floor, seat rests, seat supports, and along the overhead racks, so that in dense smoke, with eyes blinded, passengers could find their way.

Passengers who pay extra money to sit in the First Class cabin admit they like the front not just because the seats are wider and the drinks are free, but because they are not sitting over the wings which are loaded with fuel, and besides there are two door exits for escape. In the olden days of the piston planes, the rear seating areas were thought to be the safest because of the main loading door, several window exits, and the fact that the tails often broke away and spilled the rear passengers to safety.

But at the Salt Lake crash and other "survivable" disasters, the modern jets have proven that: *It Doesn't Matter Where You Sit.* The chances of escape are often based on pure luck, such as having the fuselage split open right where you are sitting.

Take the case, for instance, of the Alitalia Airlines DC-8 that crashed and burned while approaching Milan Airport on the stormy night of Friday, August 2, 1968. The pilot told investigators that his big jet was struck by a downdraft while gusty winds and wild lightning played around the aircraft. It was a situation similar to the BAC-111 crash in Nebraska. The big plane sliced into the pine trees and cut a 200-yard swath before coming to rest.

There was a lull. Ninety-five persons found themselves in brokendown pine trees and other shrubbery which appeared to have cushioned the crash. But for twelve others, the flight was their last. If they had not been fortunate enough to be where the fuselage came apart from the impact and so thrown clear, they were soon overcome by the explosion and flames that quickly followed. Carmela Fiducia was one of the victims. Her husband Amilio was thrown clear but he picked himself up and scrambled back toward the wreckage. His wife waved to him. She tossed their two-year-old son Angelo to her husband's arms. And then just as she was making her way clear she was covered in flaming kerosene and her screams joined those of the others who were trapped. She died, carrying her unborn baby with her.

The investigation of this jet disaster, number eighty-three in the Jet Decade, will probably take a year or more to complete as the disaster was another in the long series of jet calisthenics in bad storms. Fortunately there was little fuel left in the tanks as the plane was to load up at Milan for the nonstop flight to Montreal. Fortunately also the flight recorder was found reasonably intact and the crew was alive to assist in solving the mystery.

So great was the alarm felt by the industry over the mounting number of fire deaths that the CAB made a survey of just how bad the situation had become over a ten year period from 1955 to 1964. It was found that 394 air passengers had died as a result of burns in sixteen U.S. flagline crashes said to be "survivable." There may have been more victims than those located in the survey because the CAB of that day and age and the Safety Bureau of today seem to think that when an airliner slams into the ground with sufficient force to cause traumatic injuries, and when autopsies show that some indeed died of crash injuries, then the accident is deemed to be nonsurvivable. As a result of this kind of thinking, the government termed a Cincinnati approach disaster as "nonsurvivable" because of the force of the impact, despite the fact that people were trapped in their seats still alive, and were burned to death.

Therefore the figures do not show the true number of "survivable" victims. If the seats all pile into one tangle at the two bulkheads and you cannot struggle loose and flames take over, the fact that you are alive makes no difference if the sudden stop of the aircraft is of sufficient "G-force" to cause possible fatal injuries.

During the study period, which has not been reconvened since the beginning of 1965, some 153 U.S. commercial planes were involved in fiery crashes or in fire-followed accidents and of the 4,559 persons involved as passengers or crew members, 1,955 perished in the fires. This would indicate the frightening fact that over 40 percent of all passengers and crew members die in fire or fire-followed accidents

"Survivable" tragedies are created by factors that should long ago have been eliminated from modern jet transports and it seems incredible that the present fleet of jets should have been certified for use until all of the troublemakers had been eliminated: insufficient fire extinguishers, lack of exits for the trapped passengers, poor evacuation techniques, toxic-forming interior furnishings, flammable upholstery and other decorations, lack of emergency lighting, poorly outlined escape doors and windows (especially in smoke), poorly anchored seats, dangerous locations of hydraulic fluids, fuel, fuel pipes and fuel vents, and a disregard by airlines and their cabin personnel of training the travelers how to use escape windows and doors in case of fire.

Incidentally, passengers *are* told about the oxygen masks, which do not always fall down when they are supposed to. But nothing is said to the poor mother with the child on her lap about what to do when only one mask falls and she has thirteen seconds to get air to her lungs as well as to the lungs of the child . . . from that one mask.

The Safety Board also shows great concern about the "survivable" conditions of "in-flight" problems. It is believed that if sufficient and proper fire fighting equipment had been aboard the United Air Lines Flight 823, en route from Washington to Knoxville on July 9, 1964, the cabin fire might not have incapacitated the crew and thirty-nine more people would have lived. This unfortunate flight was never included in the survivable crashes because it impacted with a mountain and killed almost everyone aboard.

One passenger—who could not hope to survive the interior fire—opened his emergency window and jumped out. He died also.

This flight had been proceeding uneventfully toward Knoxville when, at 8:10 P.M., it was observed flying at a very low altitude, trailing smoke. A passenger was seen dropping from the aircraft and a short time later an emergency window was seen falling. Soon after, the aircraft flew into the ground and exploded.

Investigation of this disaster was difficult because the aircraft burned completely and the Flight Recorder not only failed to survive the impact, but its roll of aluminum film was consumed in the flames. But investigation of fire-damaged and sooted parts from every conceivable part of the aircraft led to only one logical conclusion. An uncontrollable in-flight fire of undetermined origin in the passenger compartment probably caused the crew to be overcome, or caused them to lose control of the aircraft when the passengers rushed to the rear of the plane to escape the flames.

The CAB found some interesting items. There was no smoke detection alarm other than the crew's sense of smell or observation in this aircraft. Normally, cabin air is recirculated through ventilators in the cockpit and cabin smoke or any smoke generated even under the cabin floor would be transferred to the cockpit in seconds. In the cabin only one of two CO-2 fire extinguishers had been discharged. The cabin water extinguisher had been prepared for firing, but had not been discharged. A flight crew walk-around oxygen bottle was recovered with the control valve open. One of three of the crew's full-face smoke masks was recovered but it had never been used. The forward cargo CO-2 extinguisher had been fired electrically, which brought up a new problem.

The official report said: ". . . several static and inflight tests were conducted on the fire extinguishing system for the underfloor cargo compartment . . . the results of these tests pointed out certain discrepancies which could seriously affect the safety of the aircraft and passengers. At least 15 fire extinguishers (Pyrene Duo Head Model DCD-10) were discharged. This extinguisher is located behind the First Officer on the right hand side of the flight deck companionway. At least five of the 15 tests resulted in gas escaping into the cockpit where CO-2 concentrations at head level were measured at maximum values of 12 percent. A second discrepancy concerned the improper installation of the metal seal diaphragm which is installed in the discharge head of the extinguisher. Several were found to have been installed off-center resulting in improper and incomplete ruptures.

"The Board believes that the Pyrene extinguishing system for the underfloor cargo compartment is not only inadequate for its intended purpose but also poses a danger to the flight crew."

The Board ordered corrective action. But the incident points out that modern aircraft are ill-equipped to protect passengers from fire. Take the case of the classic example of "survivable" crashes which occurred at Denver's Stapleton Field. A DC-8 jetliner swooped out of the sky, touched down smoothly on a white ribbon of concrete and sped toward disaster. Within seconds the plane swerved off the runway, flattened a truck, slammed into a new and exposed taxiway edge, burst into flames when the wing tanks ruptured, and for 16 of the 122 aboard the plane became a blazing tomb. Two others died later from their burns.

This was a flight that was experiencing hydraulic trouble en route from Omaha to Denver. As a result of this abnormal situation, fire fighting and other emergency services were patiently waiting for it to land. The passengers were told by the captain not to be alarmed by the sight of the fire and ambulance equipment along the side of the runway. The story of the disaster is important because it points out that even when informed that trouble can be expected, modern fire fighting facilities are no match for kerosene-fed flames nor for the toxic- and smoke-producing materials in the passenger compartments.

Flight 859 originated in Philadelphia on July 11, 1961, for its run to Chicago, Omaha, and Denver. At Omaha the jet was serviced with 39,000 pounds of fuel and the gross takeoff weight was calculated at 192,901 pounds, well within the limits. At 20,000 feet, the DC-8 made a slight yaw to the right and a strange tapping occurred on the control wheel. Other things began to happen and the crew decided, after referring to the DC-8 operating manuals, that they had an "abnormal" rather than an emergency situation and decided to continue on to Denver. They let the landing gear "free fall" and the big jet began its final approach to Stapleton Field at 150 knots. It crossed the runway threshold between 125 and 128 knots, which was right on the button, made a normal touchdown on the concrete at 120 knots. But then, the giant airliner lurched to the right, left the runway, struck a panel truck 300 feet off the runway, killed the driver, and slammed into the taxiway.

The second officer, anticipating an emergency evacuation, went to the passenger cabin and reached the forward loading door just as the jet came to a stop on its belly. After opening the door and installing the emergency slide, he noticed a fire burning on the left side of the aircraft. The captain and the first officer rushed to help get the passengers and jumped to the ground and held the forward escape chute. The passengers in the first class section were evacuated and now fire completely covered the front door.

Meanwhile in the tourist section at the rear, the main loading door would not open, having been damaged when the aircraft struck the panel truck. As a result, the stewardess opened the galley door and assisted as many passengers as she could away from the aircraft. But where was the fire fighting equipment? The fire engines were on the field when the jet landed and the aircraft stopped one mile from them. Eyewitness estimates of the elapsed time prior to the arrival of the first fire trucks varied from five to ten minutes and up to an incredible fifteen minutes before any effective equipment was in position.

Stapleton fire fighting crews testified their equipment was moving before the airplane came to rest and fog was being applied to the fire between one and two minutes after the crash. The Aurora fire chief said his equipment arrived at the scene eight to ten minutes after the crash and Lowry Air Base had all its emergency and rescue service at the scene in fifteen minutes—yet it took thirty minutes to bring the fire under control.

Reconstruction of the events showed that, through the forward left main loading door and the aft right galley door and from two over-the-wing exits on the right side of the cabin, 106 occupants evacuated the burning aircraft.

This was a survivable accident on a modern airport with emergency equipment alerted to an in-flight problem and yet sixteen passengers died under textbook evacuation conditions. Said Richard Martin, manager of the Stapleton Airport: "This thing focuses attention on the inadequacies of the fire fighting equipment in the U.S. There are no standards." His words could only emphasize that neither the FAA nor the Safety Board has ever set down the minimum requirements for airport fire protection and that more than half of the airports used by commercial airlines have no fire control or rescue facilities whatsoever. Only three cities in the entire U.S. have set up disaster programs for their airports.

The main reason: airports are operated by municipalities to make money. Airports must show a profit. Expensive fire fighting equipment standing by twenty-four hours a day, day in and day out, with maybe never a crash in years and years, is considered a waste of money; the Federal agencies concerned with safety, and the U.S. Congress concerned with people, have done nothing to help or force the municipalities to maintain adequate fire fighting and rescue services.

Yet, 90 percent of all air disasters occur in the vicinity of airports and more than 50 percent are fatal landing mishaps. Because of short runways and the lack of arresting gear to stop overshoots, insufficient approach concrete to allow for undershoots, and the slowness of fire equipment, lack of fire rescue training and the squeezing of mushrooming subdivisions closer and closer to airports, pilots are left with only one conclusion: U.S. airports are death traps.

In two crashes at Kennedy International Airport, one of the three airports in the U.S. with disaster programs, the fire and rescue equipment could not even find the downed airliners because of bad weather, yet bad weather usually prevails when crashes occur. In one of these, twenty-five persons died in an accident that was deemed to be survivable. It took twenty-one minutes for fire apparatus to arrive.

"Airport fire and rescue services have not kept up with

advances made in aviation," said Charles Ruby, president of the Air Line Pilots Association.

"They could never cope in the past with a DC-3 crash let alone hope to rescue more than 100 passengers from a fiery jet airliner."

Crashes in the vicinity of airports rarely show any similarity except that they burst into flames in split seconds after the breakup which usually rips seats from their moorings and hurls passengers, seats and carry-on luggage into a tangled heap that even under the ideal conditions of a rehearsal is almost impossible to untangle. Adding smoke, fire, and death-dealing fumes from interior furnishings, which should have been illegal in the first place, leaves little to assist the passengers who are "lucky" enough to survive the impact forces, and it has been proven by FAA tests that passengers can withstand tremendous impact forces.

Take the case of American Airlines Flight 383 which crashed two miles north of the Greater Cincinnati Airport on November 8, 1965, at 7:02 P.M. This Boeing 727 struck a tree-studded slope, churned onward for 120 yards and in a few seconds was shaken by a violent explosion. Other minor explosions of fuel followed and eventually the passenger cabin was destroyed by fire. Fifty-eight died.

Passengers survived this violent crash because they were spilled out of the fractured fuselage. The majority of the passengers—perhaps all of the passengers—died from the fire, according to the pathological report, though some may have died of smoke asphyxiation. The 727, like other jets in commercial service, has no water sprinkling system in the passenger compartments to force down smoke and keep the interior cool enough to help passengers escape. Doors often jam, making it impossible to escape through them. Seats break loose and seat belts trap their occupants. It isn't a pretty picture. One survivor of a recent air crash said that passengers jumped across the seats, leaping from seat to seat and from shoulder to shoulder, tramping down other passengers in their frenzy to escape the flames. No expert is required to figure how utterly impossible it would be for a mother and her children or the elderly or the crippled to escape under such conditions, which are "normal" in crashes.

At Cincinnati there was no indication that any of the doors or window exits were ever activated, reported the CAB. There was not one single passenger seat intact. One of the on-site witnesses told the Board he could look over the aircraft at one time and could see it was "full of stuff, like broken seats, stuffing from the seats and other things." The airport fire department responded to the alert. Airport truck No. II with 1,000 pounds of dry chemical was met at a highway intersection by the fire department from Hebron and the firemen from Hebron advised the airport truck that the crash was on the Dolwick farm. The Hebron and the airport fire apparatus with an ambulance proceeded to the Dolwick farm . . . *but it was the wrong Dolwick farm.* By the time they found the right location, other equipment had reached the scene. It took an incredible fifteen to forty-five minutes for all fire and rescue apparatus to reach the crash site, although it was only 1,350 feet from a paved highway.

One or two helicopters, with fire-extinguishing chemicals and their tremendous downdraft capabilities to keep fuel fires mushed along the ground, could have been at the scene in two or three minutes, but Cincinnati airports cannot afford helicopters. When the President of the United States takes off in his Boeing 707 jet, two helicopters are hovering over the takeoff area "just in case." Long ago the Air Force, the Navy, and the FAA recognized the need for whirlybirds in crash rescues. But who is going to pay for them?

The Safety Board and the FAA have also recognized the need for an airport "grid system," so that fire fighting and rescue equipment can pinpoint the crash site immediately, but only a handful of airports in the United States have ever adopted such a system. Little fire apparatus has radio control to keep in touch with the airport tower or other rescue equipment. None of them has radar, although there are cases on record when fire fighting crews were unable to find downed airliners right on their own airports.

Robert Patterson, while president of United Air Lines, was

invited to one of his company's jet evacuation demonstrations. He watched lithe young stewardesses climbing out of windows, opening doors, sliding down escape chutes, much to the amusement of others watching the tests. Patterson, never known to hold back a caustic comment and always dedicated to air safety, turned to one of his executives and said: "Invite me back . . . but next time I want to see old ladies in the aircraft, cripples, mothers with children, and then I want to see fire and smoke and confusion and then, and only then, will I be able to comment on whether or not we are able to evacuate an aircraft within two minutes or less, as we are required by law."

He was never invited back. On his retirement he received the Monsanto Award for Safety, justly awarded, and on that occasion he said:

> Safety in the air and safety on the ground must be first in everything we do. All experts are experts the day after the crash and using such expressions as "God, could they have come in the day before," is like cheating at solitaire. We are adding more seats in aircraft to keep down seat-mile costs. But which should come first—economics or safety? I have watched evacuation tests by young people who looked and acted like gymnasts. I saw young women jumping out of simulated crashed airliners like jacks-in-the-boxes and eleven of the youngsters were wearing slacks.
>
> To make it possible for survivable accidents to be survivable we must check the type of passengers in planes, seat them near exits accordingly, and make sure that before every flight all safety procedures are outlined. Someone recently said to me that accidents are a calculated risk, so why worry, Pat? My answer was, and is, that you cannot take a calculated risk with the poor soul who buys a ticket and expects you to do the right thing. I have gone to 80 percent of all United Air Line accidents and have visited vast numbers of people who have been caused hardship by these disasters. We issue between 600 and 700 pages of safety policy and no one has ever read it, and we print it like newspapers. My answer is that we must fly an airplane as if our mother and our children are in the passenger seats. Things are so bad at our airports in this country that by 1970 we will see them begin their path to obsolescence and by 1975 they will be completely obsolete. The larger planes they are now planning should be for cargo. I

don't think in the interest of safety that these larger jets are ready for passengers yet, considering safety performances at our airports.

Captain J. W. Meek, of Delta Air Lines, a member of the Airport Problems Panel of the Air Line Pilots Association told an air safety forum: "There has never been an item that affects the performance of an airplane added for purely safety reasons."

Said Captain B. V. Hewes, also of DAL and chairman of the ALPA Rescue and Fire Committee: "A large majority of major airports have vastly improved their fire and rescue services . . . unfortunately a large majority of medium and smaller airports have made absolutely no progress and local service airlines are still operating with little or no protection." And now they have jets using their fields.

Captain Hewes told the Air Safety Forum that studies by his group show that most crashes that occur in the airport areas are survivable.

Survivability time is reduced to a matter of seconds where there is a disintegration of the fuselage or where all the emergency exits have been opened to the fire area. New methods of fuselage construction to withstand impact damage are needed but let us not forget the obstacles themselves. We need grading of small hills, filling in holes, cutting down trees and most important of all, building decent overrun areas.

As a further aid to reducing the fire hazard we need ignition suppression. There are many ignition sources during an aircraft crash, friction sparks, static discharge, hot engines and electronic equipment, to name a few. For the suppression of friction sparking in the past we have advocated runway foaming but with the increased use of titanium, a metal that produces a very intense spark, a foamed runway is going to provide us with little or no help. Inerting of an aircraft engine is practical, but very costly in weight and dollars and for this reason development seems to be at a halt. We must look somewhere else. Therefore new methods of fuel containment must be sought as soon as possible—bladder tanks, stronger tank construction and so on.

Captain Hewes said that emergency exit markings are good in daylight, but completely inadequate during the dark hours or when a fuselage is covered with dirt or foam. He advocates the use of strobe lights in all windows of emergency exits. He added:

> And now we have a new problem. How do emergency crews locate the scene of an accident with the new instrument minimums proposed by the FAA. With the proposed zero-zero landings, the chances of an airport accident are greatly increased.
>
> If the pilot cannot see far enough ahead to stay on the runway, just how does the fireman see to drive his fire truck, even if he was previously informed of the crash location? How does he find his way out there? Our new exit light would help but what we need is a radio locater beacon on the airline and [to] equip the fire trucks with a receiver.
>
> Our committee is still pressing strongly for increased research in the field of survivability . . . BUT WE HAVE A LONG WAY TO GO.

Captain Tony Spooner, former BOAC skipper and Technical Secretary of ICAO, stated that if present accident rates remain, 60,000 air passengers per year will be killed in thirty to forty years' time. Captain Sidney Hill of Western Airlines said: "The industry has made its greatest strides during the past 40 years in dealing with those problems of aviation which occur above the runway height . . . but from the point where the glide path passes over the 1,000 foot mark down to the end of the runway, the industry has failed to keep pace with the aircraft's needs, and only with comparable progress from this point on down can we anticipate achieving an acceptable level of airport safety."

Pilots claim the means for preventing accidents at airports are known. The time to start improvements was yesterday. Little is being accomplished today. A National Airport Survey conducted by the Air Operators Council disclosed that "Five billion dollars would buy adequacy."

A wide range of safety devices to improve evacuation from crowded jets and to assist in the survivability rate was described at the FAA-industry crashworthiness conference on March 29 and 30, 1966. More than two hundred participants took part in the comments and recommendations—an indication of the scope of the seminar. A report on emergency evacuation demonstration by R. B. Stophlet, FAA Task Force Member, revealed that in 250 simulated aborted takeoffs and gear-up crash landings there were seventy "deficiencies," fifty-nine of which involved evacuation slides. Slide problems were caused by design, improper maintenance, lack of crew training. The other deficiencies involved such things as: stuck cabin door, jammed overwing exit, exits blocked by malfunctioning jump seats, and inadequate crew training.

Robert McGuire, of the FAA's Aircraft Development Service reported on research and development in two broad areas: crashworthy structural design and minimization of fire hazards. Specific projects now underway by the FAA include crashworthy structural demonstrations, crash resistant fuel tanks (both bladder and integral types), sparking characteristics of structural materials, dynamic testing of cabin seats, combustible characteristics of interior materials, the use of shaped explosive charges to create additional emergency exits, use of cabin water-fog or detergent-foam systems, and techniques to improve exit locating.

McGuire said his agency is interested in a fog system that would use an airplane's water supply and dispense it through a cabin sprinkler system. In Detroit the Metropolitan Airport fire department was demonstrating a foam system before a group of spectators. One fireman disappeared into the foam and failed to come out. He had slipped on the foam and died from the concussion.

Explosive charges would be created of two chemicals separated from each other and contained in the shape of a tube around the fuselage area to be blown open. A switch or rupture would bring them together and the side would blow outward. McGuire said the FAA is also interested in an external explosive system so that rescue crews could force their way into an aircraft and not have to depend on inadequate axes, as in the hopeless case at Salt Lake City.

Thomas Horeff, Propulsion Program Manager of the FAA, said that gelled fuels offer some promise in the continuing quest to find some way to stop the explosiveness and spreading of fuel. Gels, or thickened fuels as they are sometimes called, have been under test for four years and show some promise. Gelled fuels burn more slowly and less intensely than unthickened fuels, but the main benefit is the slowness in which this type of fuel oozes from ruptured wings. It doesn't flow like kerosene or gasoline or float to the top of water to compound the hazard. The FAA is also expected to issue research proposals soon for investigation of fuels that would change to solids under impact forces.

Among many presentations was an interesting technique shown by Dr. Stanley Mohler, chief of the Aeromedical Applications Division, whereby airplane passengers would wear a transparent hat over their heads during emergency evacuation and the hat, which would be fire resistant, would trap air and allow passengers to get clear of the aircraft without burning out their lungs. He also suggested that a new look be taken at aircraft interiors to eliminate all hazards such as sharp points, brittle objects, and instrument knobs. All passengers and crew members should wear shoulder strap harnesses that would withstand 25 Gs of force, he said.

Air Line Pilots informed the Crashworthiness Council that they wanted wider aisles in all jet transports, more emergency exits, rigid evacuation chutes, emergency lighting and marking, a fully trained attendant at all exits, elimination of carry-on baggage—which has interfered with evacuation in the past—a cockpit door that the crew can break through in emergencies, self-contained fire-extinguishing systems, fireproof cabin materials, and greater fire protection in cargo departments. ALPA also asked that flight recorders be ejectable, locatable, and floatable.

"This seems criminal to be advocating all these necessary safety innovations in modern jets at this time when they should have been done before the jets were ever introduced into service," said James Foy, until recently the president of the International Federation of Air Line Pilots. "I fly a DC-8. There are only eight passenger escape exits and that is not enough. If two exits are provided for the Flight Deck crew, how many more should be required for the 130 passengers and more who are in the cabin behind us? Our airport fire and rescue services are out of date and since none of them can reach a downed jet in two minutes, every aircraft must carry its own fire fighting system.

"In the old days we crashed into mountains. Now we crash at airports because most airports are unsafe and we have therefore established a new jet age phrase 'survivable accident.' At least the mountains were merciful," he said.

8

The 727 Story

The Boeing 727 jet airliner made its debut with all the fanfare of a Broadway "first nighter." Billed as America's first short-to-medium-haul jet this three-engine transport rolled from Boeing's factory at Renton, Washington, in November 1962. It was the first of 127 aircraft that had been eagerly purchased by United Air Lines, American, TWA, Lufthansa, and Trans-Australian. The tri-jet joined the more than 300 Boeing 707 and 720B jets in world service with twenty-eight airlines and the more than 550 Boeing jet transports and tankers in the United States Air Force.

The introduction of the 707s and the DC-8s had shrunk the world by 40 percent. The short-to-medium-range jets promised a new era in transportation between smaller cities, which were now to experience the noise and kerosene fumes of modern flying. The 727 had a lot of scamper, could climb quickly and descend at 2,000 feet a minute, handle itself in turbulence better than its bigger brothers (no one could see the engines bouncing on their pylons), and introduced a new degree in quietness to the First Class Passenger section, which was ahead of the sweptback wing. Generally, the flights were so breathtakingly different that passengers were plotting their timetables so that they could fly the 727.

Unfortunately, like many new turbine-powered commercial aircraft, particularly the rear-mounted engine type, the Boeing 727's record was marred by a series of terrible disasters, all of which occurred at night during approaches to big-city airports. Within seven months, four of the giant airliners had been destroyed while in use with major airlines, and an incredible number of innocent victims had been blown or burned into eternity.

Four disasters between August 16, 1965, and February 3, 1966, claimed the lives of 263 passengers and crew members, and travelers began to take a second look at the attributes of the Boeing jet. United States insurance companies listed three of the crashes as catastrophic, because they occurred in the United States and cost the companies more than $15,000,000 in claims. Airport-bought insurance on these disasters cost one insurance company almost $2,000,000 in payments to beneficiaries. The loss of the jets totaled more than $24,000,000, and the lawsuits that were emerging were enough to severely scare the airlines and the manufacturer.

Disasters like these had happened before with the Electras, the Boeing 707s, the Douglas DC-8s, the Comets, the Martins, the BAC-111s . . . but they were nothing in comparison with the 727's disasters in frequency, pattern, or loss of human life. Although the CAB and the FAA argued that there were no similarities in the crashes, and that there were no reasons for grounding the jets, the four disasters left air travelers apprehensive. Only the years ahead would erase these emotions.

True, there had been some unusual procedures in the final approaches to airports by two of the 727s—at Cincinnati and at Salt Lake City—and this gave rise to the feeling that the pilots might have been partly to blame. Perhaps they were. But it seemed strange that veteran multimillion-mile captains, with thousands of hours in jets under their belts, were 100 per cent to blame for landing troubles in a new jet aircraft, when four such planes crashed in such short order. And although the government agencies could argue that the disasters were unalike, simple arithmetic seemed to indicate a pattern between the jet that fell into Lake Michigan off the Illinois shore in August 1965 and the Nipponese 727 that soon afterwards fell into Tokyo bay on its final approach to the Tokyo airport.

With so many people attempting to place the full responsibility on the pilots (one was living and could tell his own story), it was logical that the pilots would take a strong stand. As could be expected, the CAB and the FAA began to receive reports, persistent reports, that pilots were experiencing rapid rates of sink during 727 approaches to airports as well as during other periods where rapid decreases in forward speed were necessary.

What to do? The 727 was in service on a score of the world's airlines. Could it be that pilot error was the cause of all of the crashes, or three of the disasters, or perhaps just two? Time alone would tell—the time it took to learn the secrets of the flight recorders, if they could all be found. Meanwhile, new operating procedures for the 727s were being issued and beefed-up pilot training for the rear-engine jets was quietly underway on all the airlines with planes of this type.

To have grounded the 727s at this stage of the investigations might have been foolhardy, since there was a chance that pilot error could have been a factor in all cases, and the 727 was still first choice with passengers.

And no wonder. The aircraft was capable of slicing through the thin atmosphere at 600 m.p.h. with up to 114 passengers over distances ranging from 150 to more than 2,000 miles. Exactly one year after it entered commercial service on February 9, 1963, there were orders for 230 more clogging the Boeing sales books.

The 727 was different from all other rear-engine jets, because it had three power plants, not two or four like the others, and it boasted a fantastically high tailpiece, almost thirty-four feet from the tip to the ground. Two of the engines were mounted in pods on either side of the rear fuselage, with the intakes close to the last window on each side of the aircraft. The third engine was cowl-enclosed and suspended from a beam in the tailpiece.

The 727 was approximately 134 feet long, with a wing span of 108 feet, seven inches, with a sweepback of thirty-two degrees at the quarter chord line. Triple slots in the wing-trailing edge flaps and leading edge flaps and slats provided the 727 with unprecedented high-lift capability, enabling it to operate from 5,000 foot runways at gross weights of 152,000 pounds. With this wing design, and with the three power plants boiling out the superheated air, it could take off in short order with 26,000 pounds of payload, and with 7,000 United States gallons of kerosene in its thin wings, it could climb to its best cruising altitude of 35,000 feet in less than twenty minutes. Actually, the 727 was designed to operate economically between 15,000 and 35,000 feet, while its operational ceiling was rated at 42,000 feet, the same as most other passenger jets.

The 727 pleased the airlines, most of the pilots, and the long-suffering public. Boeing quickly announced that a "stretched" version of this jet, to be called the 727-200, would be introduced by late 1967, with an all-up weight of 169,000 pounds, maintaining the same speed with the same engines. When this version was finally introduced to service in 1968, it was immediately banned at Washington National Airport as no longer a medium-short-haul jet.

Those who experienced the takeoff of the 727 never quite got over the rocketlike swoosh as it roared down the runway for what seemed a very short distance, and then climbed at a high degree from the concrete below. The front section was very quiet and from the windows it was almost impossible to see the wings let alone the engines, which were hidden completely from view. The Tourist Section over the wings was considerably noisier.

There were some passengers—and there will always be those who can never be pleased—who worried about engines mounted in the rear. They were apprehensive that in case of a severe crash-landing or a sudden stop against a boundary fence, the engines would travel into the passenger cabin. Some engineers did not like rear-engine mounting, fearing that the catastrophic explosion of an engine or the loss of a turbine blade could shear off the tailpiece, the power controls, as well as rupturing fuel lines. Also, passengers were not only seated over the fuel of the wings, but now they were also above the high-pressure fuel lines leading from the wing tanks rearward to the engines. The design of the 727 allowed a great deal of fuel to be present in close proximity to the passenger cabin, with fuel lines that could be ruptured in case of wheels-up landing.

So there was good and bad about the 727. Then on the stormy night of Monday, August 16, 1965, the first Boeing 727 went down with twenty-four passengers and a crew of six. One of the passengers was Clarence "Clancy" N. Sayen, who was President of the Air Line Pilots Association from 1951 until 1962. It was an ironic turn of fate for the man who had fought long and hard for aviation safety to become a statistic on an ill-fated journey.

The flight, operated by United Air Lines as Flight 389 from New York's La Guardia Airport to O'Hare International Airport in Chicago, departed at 7:52 P.M. It was due in Chicago at 9:50 P.M. but en route it had made excellent time, and it was coming down through scattered clouds over Lake Michigan when the flight called the O'Hare Approach Controller. At 9:18 P.M. the 727 was given its landing pattern. The radar blip of the aircraft was clear as it descended gradually over the lake toward the Illinois shoreline about fifteen miles ahead. The cloud base was at 7,000 feet, and the aircraft slipped below it about 9:19 P.M. without any report of turbulence or malfunctioning of the engines or controls. The last report was "Roger" from the copilot when the landing instruction had been given.

Ahead of the aircraft now was the shoreline of lights and to the southwest the brilliant glow of Chicago. Ahead and slightly to the southwest was the O'Hare beacon. Visibility was good. A few lightning streaks played through the clouds beyond the south end of the lake from the cold front, which had passed through the area earlier and was nasty enough to stir up one or two tornadoes. Still, there was no hint of turbulence from the airliner and there were no storms in the vicinity.

Captain Melville W. Toule, forty-two, of Wyckoff, New Jersey, was at the controls with First Officer Roger Wipezell, thirty-four, of North Plainfield, New Jersey, at his right hand, and Second Officer Maurice Femmer, twenty-six, of Elmont, New York, behind him at the engineering panel. The three stewardesses aboard this flight were Phyllis Rickert, twenty-two, of Mount Prospect, Illinois, her roommate Sandra Fuhrer, twenty, of Long Island, and Jeaneal Beaver, twenty, of Long Beach, California. Donna Williams who shared the apartment with the first two girls was supposed to have been the fourth stewardess on the flight but at the last moment she had canceled out.

Approximately two minutes after the last report from the flight, two things happened. The aircraft disappeared from the radar at O'Hare Field and a lifeguard at North Avenue Beach, southwest of the plane's position, saw an orange flash over the lake and seconds later heard a thunderous roar. The time was about 9:21 P.M. Baggage carts for the flight had left the terminal to meet the plane, but someone at the airport had seen the flash in the sky and rumors quickly spread that there had been a disaster, a chilling thought to those lonely ones awaiting the arrival of the 727.

Had witnesses seen the flash in the sky? Or had they seen the flash reflected on the clouds, a flash caused by the explosion of the jet when it struck the water? The lifeguard had heard only one explosion, so it seemed logical that the jet had blown up only when it hit Lake Michigan's cold surface. Had events occurred so swiftly that the pilots had no time to radio? Was it a bomb? Was it weather? Did the pilots see another plane ahead and, in trying to avoid it, lose their bearings and their control of the jet?

The flight recorder would tell the story. By superb skill and back-breaking work, using nets and grappling equipment, underwater television and sonar over a period of months in high waves and bitter winds, a group of dedicated men dragged to the surface about 81 percent of the wreckage, but not the recorder—although its outer shell was found. They retrieved all the bodies, usually in nets. The cold waters had preserved the victims as if they had just died. The autopsies on the crew and passengers proved that every death was caused by traumatic impact. Most died from fractures of the neck. In other words, the passengers were alive until the aircraft struck the water and exploded. Many were cut in half by the safety belts. There was no indication whatsoever that a bomb had been on board. No residues were found in the lungs, flesh, or clothing of the victims that would indicate an explosion. In fact, the appearance of the wreckage seemed to indicate that the aircraft had struck the water in an attitude comparable to a normal landing approach. One observer at the scene said it looked as if the giant plane had made a perfect landing . . . on the lake. It had all the earmarks of a first-class mystery. And the cause was never found, though it appeared that a misreading of the altimeter by the crew might have been the culprit.

Then, three other 727s crashed, all while approaching airports. Number two in the chain occurred when an American Airlines 727 Astrojet was approaching Greater Cincinnati airport on the blustery rainy night of November 8, 1965. Fifty-eight persons died in the fiery wreckage when the aircraft ran into a hill.

Like the flight that had ended in Lake Michigan, this 727 service had started at La Guardia Airport. The airliner flew directly to the base leg of the Cincinnati airport and then made a turn south in order to land on Runway 18—the number given to this runway because the south landing is made at 180 degrees. A witness described the jet as "very low" while over the base leg, and crossing above a college.

The wind was between eight and fifteen knots from the south, and showers, some heavy, were falling at the time. But the airport was visible from the airliner. In fact, visibility had been ascertained at from six to eight miles, and although the flight had been on Instrument Flight Rules on this nonstop trip from New York, Captain Daniel Teelin, forty-six, elected to cancel the IFR plan to make a visual approach to the airport. He would not have done so unless the airport was in the clear. Not only was the surface visibility excellent, but the cloud ceiling was up at 1,500 feet. There was an occasional heavy shower on the approach path, and showers often mean turbulence and sometimes downdraft.

Two miles short of the threshold of the runway, the 727 crashed into a wooded slope about a hundred feet below the summit of a hill that rises some 830 feet above the Ohio River.

Witnesses to the crash said they could see the jet was not going to make the hill. One woman spectator screamed before the collision.

What had happened? Too many persons were too quick to say there was no similarity between the Lake Michigan and the Cincinnati crashes. Captain Teelin was a top pilot, a man who would not be expected to fly his jet into the ground. He had twenty years with American and was one of the carrier's superintendents of flying at La Guardia. He had amassed 16,000 hours of flight time, of which only 300 hours were in the 727s, not much to ascertain the jet's possible idiosyncrasies. Yet, it certainly looked as if he had flown into the hill. Did his aircraft sink suddenly, sink almost horizontally? The flight recorder disclosed the aircraft increased its vertical speed from 850 feet per minute to 2,100 feet per minute—about fifteen seconds before the crash, although in the last ten seconds its airspeed fell from 160 to 150 knots. The recorder also disclosed the jet was in turbulence during the last fifty-five seconds of its life.

The copilot was Captain William J. O'Neill, thirty-nine, who had logged more than 14,200 hours in his fourteen years with American. At the time of his death, he had flown 2,500 hours on jets but only thirty-five hours in the 727s. The flight engineer was John T. LaVoie, thirty-three, a ten-year veteran of American. All those on the flight deck were leaders in the business.

One of the passengers who miraculously survived the holocaust was veteran American pilot, Captain Elmer Weekly, forty-six, who was an important witness in the crash investigation. Also surviving was a stewardess, Toni Ketchell, twenty-five, and two passengers, both of whom were seated back in the rear coach area, having bought Tourist tickets for the flight. All the First Class passengers died, proving once again that it doesn't matter where you sit.

Investigation revealed that five seconds before the impact the crew, in communication with the tower, had no indication whatsoever of the jet's low altitude. The Safety Board determined, after sifting all the facts and after a great deal of criticism from the Air Line Pilots Association, that the probable cause of the accident was the failure of the crew to monitor the altimeters properly during the visual approach into deteriorating visibility conditions.

Coupled to the findings was the following notation by the Board:

> The Board would be remiss if it did not take cognizance of the concern that existed in the minds of many elements of the aviation community and traveling public regarding Boeing 727 crashes . . . investigation did uncover certain aspects of operating practices in the Boeing 727 that warrant industry attention in the interest of preventing future accidents . . . it was noted that close-in, high descent-rate, unstabilized approaches are being conducted more often in the 727 than in any other jet transports studied . . . why this is true is not evident from the preliminary review and any realistic evaluation will have to wait until a final NASA report on this matter.

The Board thought also that "consideration must be given to the fact that the 727 does have highly responsible and versatile flight characteristics [which] may be misleading to the pilot or are presenting the impression that greater liberties may be taken . . . "

Then following on the heels of the American mishap, another United Air Lines 727 crashed on its final approach to the airport at Salt Lake City, Utah. The jet, gliding rapidly down to the runway touched short, bounced, skidded, and twisted into a fiery nightmare that took the lives of forty of the ninety aboard. This disaster has been described fully in an earlier chapter.

Public relations typewriters were busy grinding out the number of passenger miles of safe operation that had been chalked up by the airlines. But statistics could not bring back the dead, nor could the imaginative promoters of the airlines dim the fact that too many people were dying, and others were being exposed to death and injury, in modern jet aircraft. More people were being killed in jets than were ever killed in piston aircraft, and the disasters were not confined to this country.

On the heels of the three 727 crashes in the United States came the worst air disaster in all commercial history. Again the 727 was involved, and the circumstances appeared strangely like those of the Lake Michigan crash. All-Nippon had made three jet flights on February 3, 1966, shuttling hundreds of government officials, Tokyo executives, and tourists to the seventeenth annual Snow Festival at Sapporo on the western coast of Hokkaido, the northernmost island in the Japanese group. Hundreds of thousands of visitors had thus far thronged to the Festival to observe the hundred-foot-high statues of ice. This year's exhibit had been the most ambitious in the island's history, because civic officials were anxious to have their city chosen as the site of the 1972 Olympics.

On its third and last shuttle for the night, All-Nippon cleared Chitose Airport and headed for Tokyo and the 300 mile hop that would take less than an hour. On board were 126 happy, laughing travelers, many of them top Japanese executives and industrialists, four stewardesses, and three crew members.

At 7:00 P.M., the big jet was cleared to the Tokyo Airport. It descended through light showers for the final approach, which was over Tokyo Bay to the airport on the shoreline. The pilot reported an instrument malfunction, but no emergency was declared, and he elected to come in with a normal approach involving a turn over the bay and then directly into the active runway. Six miles from the threshold and approximately two minutes from touchdown, something went wrong. The jet fell into the choppy bay, and, at the same moment that the radar controller was giving the emergency alert to rescue services, a fisherman saw a red fireball erupt over the water and rise like an atomic bomb to the rainclouds above. There were some who thought the explosion had first occurred in the sky, but most witnesses agreed it happened when the 727 hit the surface and that what the others saw was the reflection on the clouds, a situation similar to the Lake Michigan disaster.

The loss of 133 persons stunned the Japanese nation. At the Tokyo Airport, hundreds of screaming women and children and groups of weeping men milled in confusion as the news was released that all the passengers had died. The Japanese government promised a full inquiry, the Boeing Company sent three top experts to assist in the investigation, and the FAA again reiterated that the 727s would not be grounded, as there did not appear to be any similarity in the crashes—except, of course, that they were all on approaches and all at night, and two over water, which the reader might think is reasonably similar.

With an all-weather goal in mind, the 727 had been built with high-lift devices for lower landing-approach speeds, a new autopilot, and a change in control systems. The philosophy behind the Boeing approach was that the human pilot should always remain in control of the aircraft while the aircraft itself would be given added capability for safe, low-speed operation in poor weather. As an added engineering plus, irreversible hydraulic controls on all axes provided the needed precision in controlling the attitude of the jet during low speed in bad weather. As a further aid, the 727 had a unique windshield rain-removal device in conjunction with newly developed high-speed windshield wipers which improved visibility dramatically even in torrential downpours.

Weather, however, did not seem to be the cause of the All-Nippon crash, although it was raining at the time. Three bodies were recovered from the water five hours after the crash by a lighthouse ship. A book of lifesaving instructions was found nearby. A military patrol boat found part of a wing section. The next day thirty bodies were recovered while a Japanese destroyer hauled to the surface a large section of the broken fuselage. Japan plunged into national mourning, and two small coastal freighters ploughed through choppy waters with hundreds of floral tributes that were scattered over the watery grave by scores of weeping relatives.

Faced with multimillion-dollar lawsuits over the series of 727 crashes, everyone in the airlines, the government, and the industry clammed up. What use had been the great research efforts, the years of testing, the safety seminars, escape drills, computer analysis, radar guidance, navigation, pilot know-how, and the much-touted fire tests of the FAA in New Mexico, which had received so much publicity?

Insurance companies, hit with the material loss of the jets, also paid out some $13,000,000 in life and accident insurance policies from the three domestic jet crashes. From airport-bought insurance, one company paid out $1,300,000 to beneficiaries. The insurance companies began to goad safety organizations into greater activity by a threat of greatly increased premiums.

Both the airlines and Boeing moved into action concerning flying procedures because passenger concern over the 727s was mounting. The FAA had received many complaints that 727 landings were often hard landings, and travelers feared that they would end up like many of the Salt Lake passengers. Air travelers were complaining by letter and telephone that pilots were jockeying the throttles excessively during landing approaches.

"The landings are no harder than those of the larger jets," said an official of the FAA. "But the 727 has a somewhat stiffer gear than heavier aircraft, and to a passenger the touchdown may seem harder. But some people swear to us that the engines were almost knocked off by landing impact. However, the 727 pilot was not doing anything that one on a 707 or a DC-8 does not do as well."

This all may be reassuring, particularly to an official discussing aircraft approaches. But passengers were noticing the power fluctuations in the 727s because of the rear-mounted engines. The power settings and the vibrations of change were more audible in the passenger cabin, and no explanation was coming from the flight deck for these surges. Some passengers complained they were alarmed at the high increases of thrust when the forty degrees of flap was extended for the final landing flareout seconds before touchdown.

It was abundantly apparent that passengers were paying more attention to surges in power during their flights in 727s than in any other jet aircraft, and good public relations between the flight deck and the passenger section could have easily eliminated some of this worry. "The Electra is my favorite plane" was being heard more and more at airline counters—and in the jet age at that.

Captain Paul Sonderlind, director of flight standards for Northwest Airlines, said that the 727 was every bit as honest an airplane as he had ever flown. "There is no basic difference between it and many other jets in the way it handles, and in many respects it is easier to fly," he said.

Six days after the All-Nippon crash, a TWA 727 made a spectacular emergency landing at Denver. No one was hurt and the pilot was alive and could not be blamed for this mishap. In fact, his skill saved the lives of seventy passengers and the six fellow crew members.

Captain Frank L. Smith was at the controls at the aircraft as it homed in on Denver airport. Suddenly, a faulty electric motor in the automatic pilot system jammed the horizontal stabilizer on the tail, which controls the nose-up and nose-down attitude of the aircraft. Smith radioed the emergency to the tower and then told the passengers of the difficulty. Stewardesses rushed from seat to seat, removing spectacles from pockets, placing pens and pencils on the floor, tightening seat belts, and placing pillows and blankets over the heads of the passengers to save them from possible serious injury in case of a hard crash landing.

Assured that the passengers were ready for the landing attempt, Smith took the 727 into a long shallow turn and inched his giant jet lower and lower over areas that were solid with mountains. Then he placed the jet almost on the deck (just above the ground surface) and nursed it into the airport in what was described by Charles F. Stacy, FAA Flight Standards chief at Denver, as a "tremendous job." Another pilot, flying as a passenger on the flight said: "He made a landing that was damned near impossible."

It was only natural in the wake of all the disasters and the near misses that there would be demands for an investigation. The public deserved it, and surely those who had bought the 727 had the right to an explanation. They had lost a record number of passengers, and there was pressure upon the airlines for a quick answer. The FAA stubbornly continued to claim there was nothing wrong with the jets, no similarity in the disasters, and that all the fuss was unwarranted. They steadfastly refused demands to ground the jets, well aware both of the economic impact that grounding could cause and the demoralization of airline traffic.

Then, on Tuesday, February 15, 1966, Representative Henry Gonzalez, of San Antonio, Texas, called for a grounding of the jets and charged that the CAB's investigation into the crashes was at a "leisurely pace." Gonzalez was supported by Senator Vance Hartke of Indiana, who thought it imperative to ground the 727s because of the similarity of the crashes,

This forced the FAA to move. A parley was called of every airline in the world to meet in Washington to hear the reports of preliminary Federal investigation. The meeting was to smooth the troubled waters. As long as no other 727s crashed in the near future, the meeting would be able to calm the fears of airlines which had invested millions in the American-made jets. The future of the plane was at stake.

At the moment that the closed meeting took place, 226 of the jets were flying on world routes and 245 others were on order. Since the CAB and the FAA could find no reason to ground the jets—because the investigations so far had emphasized pilot errors or pilot unfamiliarity with the rapid sink rate of the 727—the meeting reassured the fourteen world airlines of the benefits of the 727, and impressed on them that flying procedures for the aircraft had to be "standardized for prescribed techniques" during landing periods.

It was pointed out that landings could be made safely with descents of 2,000 feet per minute, if pilots were aware of the precise moment they had to apply power to reduce the descent rate for a gradual approach. Investigation had shown rapid descents were made at Salt Lake City and over Tokyo Bay. The only way to stop the sink rate of such rear-engined craft was to apply power at the precise second to increase the forward speed. Yet pilots had pointed out that it could take seven or eight seconds to increase a 727's idle to full power. In a jet descending rapidly earthward, seven or eight seconds is a long time.

The FAA informed the worried operators that it had already taken steps to have pilots call out their descent rate and speeds at various altitude levels on their approaches to airports. Furthermore a bulletin issued to operators and to the Air Transport Association specified the rate of descent limits at lower altitudes; these limits would insure adequate margins for recovery. Many thought it strange that an agency charged with the duty to know all these facts before the aircraft was granted a certificate of airworthiness should issue such instructions only after 264 lives had been lost.

At this junction of the "clear-the-727" parley, the Air Line Pilots Association lashed out at the FAA, charging the agency had "prematurely" issued statements on the Salt Lake City investigation, which was not complete, and that the widely publicized statements concerning cockpit procedures might infer that the pilots were to blame.

FAA did indeed discuss the crew behavior in the Salt Lake accident, in an attempt to prove the crash was not the fault of the aircraft. Despite ALPA protests, the government charged there was "uncertainty on the flight deck as to who was actually flying the 727 during the last 1,000 feet of its descent."

It was necessary now to reveal evidence of these last moments and the hearing was provided with the following sworn testimony and investigation data by the FAA even though the agency may have been guilty of not providing sufficient information on the air traffic routing into Salt Lake City during that time. The following information was revealed:

1. The FAA air-traffic controller assigned the flight on an approach routing adopted only four days before the accident. Neither of the pilots had ever heard of the routing, which required that, prior to beginning its final descent, the aircraft must be 1,000 feet higher and between six and seven miles closer to the airport than the former route.

2. Both pilots claimed that the engines on the aircraft failed

to respond when the power levers were moved forward during the latter phases of the approach.

3. Both pilots reported that less than normal elevator control was available when they attempted to flare for the landing. (The flareout is the moment when the aircraft is approaching the runway and when, a few seconds from touchdown, the pilot pulls back on the wheel, raising the elevator slightly. This moves the nose up so that the main landing-gear wheels will touch the runway first, saving contact with the nose wheel at a time when all the weight is on the main landing gear.)

4. Data taken from the tape of the Flight Recorder showed that the 727 approached the landing with a rate of descent of 2,000 feet per minute and had held that high rate for 1.5 minutes before crunching into the concrete.

5. The pilot in command had revealed to the inquiry in Salt Lake that his First Officer was flying the aircraft during the final approach under "my verbal directions" but then when the aircraft was between 1,200 and 1,000 feet from the ground, the pilot-in-command realized that the aircraft was descending too fast and seized the control column. At that same time the First Officer thought he was still flying the plane. He reached out to apply more power and found that his captain had already done so.

It is interesting to know, at this point, that United Air Lines regulations call for the pilot who is not flying the aircraft to call out the altitude and airspeed under 1,000 feet of altitude and he is also required to call out the rate of descent under 500 feet . . . almost second by second. On the night of the Salt Lake crash, because both pilots thought they were flying the aircraft, each waited for the other to call out the speeds and altitude. Neither one did. Such was the confusion that supported the FAA's argument that the 727 was not in itself the troublemaker.

Carl Christenson, assistant vice-president for flight operations of United Air Lines, revealed that the pilot-failure rate with 727 was higher than normal. He said that of the 380 flight officers who had entered 727 training since November

1963 twenty-one had withdrawn or terminated, but all the rest had qualified. On the 727 oral examinations, forty-seven failed the first test and nine the second test. On actual flight tests, eighty-six pilots failed the first test, sixteen the second, and four the third. He said that United insisted on their pilots making a "stabilized approach," meaning that the plane should be positioned somewhere between the outer marker (5.6 miles) and the middle marker (four miles) at 500 feet, which would give the jet a graduated letdown to touch the runway at 1,000 feet inside the runway beginning line. Asked if the Salt Lake 727 was on such a specified approach, he replied: "It was not."

One of the most important witnesses—and the one to whom the airlines were paying particular attention—was George S. Moore, FAA's chief of certification of aircraft. He testified that the 727 was certified for air-carrier use in December 1963, but only after tests involving 1,200 hours of flight and 33,000 man hours over a forty-three month testing period. A review of this long test program, he said, indicated nothing was wrong with the design or flight characteristics of the aircraft.

He said that exhaustive rechecks and retests had been made following the four 727 disasters, including a simulation of the Cincinnati, Lake Michigan, and Japanese crashes as closely as the last minutes of the flight could be duplicated. In every case, the aircraft responded as it did in the certification process. Not only that, he said, but twenty-four domestic and foreign airlines, operating 226 of the 727s, had been closely probed about the 727 flight characteristics as they occurred in actual practice, and nothing significant had come to light. If there was any defect in the plane, it had not been disclosed during certification.

N. H. Pomeroy, manager of turbine engines for United, spoke of operational experience with the 727 engines. He said that in 1965 twenty-four incidents of engine malfunctioning had appeared in the logs of the pilots, and one of the incidents was a reported loss of power during an approach into Reno, Nevada. However, in this incident, corrective action was taken by the crew, and the aircraft landed with normal power coming from all three engines.

Many witnesses testified as to training, engine behavior, and stability of the jet and finally the 727 was "cleared" by the FAA as "airworthy."

Right after this, an Eastern Air Lines 727 made a spectacular wheels-up landing at Miami International Airport. In far-off Australia, an engine began breaking up immediately after a Trans-Australian Air Lines 727 took off, but the pilot wheeled the jet in a tight turn and quickly landed.

Then, in mid-1966, a special Civil Aeronautics Board report exonerated the 727, making it clear that the aircraft had no serious weaknesses or any mysterious "bugs." Pilots, the board said, liked the plane, some going so far as to say that it was the best they had ever flown. The report conceded that the 727 had a rapid rate of sink with the flaps set in a landing-approach position, but they emphasized that while this characteristic was well known it was not stressed in pilot training because pilots were not supposed to bring the planes down so fast.

The official investigation of all the 727 crashes seemed to vindicate the action of the FAA in not grounding the jets when the Cincinnati, Chicago, and Salt Lake crashes occurred. It appeared that most of the problems were occasioned by crews during approaches and at night, that the aircraft itself was a good aircraft, and familiarization with its behavior in flight was not of sufficient duration. As Charles Ruby put it: "Although aircraft have become more complicated, sophisticated, and critical in certain flight regimes, they are often introduced in mass service so rapidly that many of the things which pilots are required to know to fly them, do not make themselves evident until an adequate experience level is reached on the aircraft in actual service."

And that ended that—the 727 was cleared.

Then in the space of thirteen days in 1969, a 727 crashed at Gatwick, England, during an approach, and a United Air Lines 727 plunged into the ocean off Los Angeles. The toll: 88 dead.

9

Jet Roulette

It would be unfair for the reader to get the impression that one jet caused more trouble than others in this first decade of jet calisthenics.

From 1959 until the end of 1968, twenty-eight jets flown by United States airlines were involved in fatal accidents and of these, five were Boeing 727s, eleven were Boeing 707-720Bs, four were Douglas DC-8s, three were Douglas DC-9s, two were BAC 111s, two were Convair 880s and one was a Sud Caravelle. The best record was made by the Convair 990 which never had a fatal crash.

It was never proven to everyone's satisfaction that the cluster of 727 crashes was wholly the fault of the pilots exclusive of other factors. However, the entire question of how human beings can cope with the problems of jet airliners is a serious one. Some near-disasters have been averted by the extraordinary skill of the jet crews. In other cases, it would appear that pilots, even with other skilled flying supervisors on the flight deck with them, made incorrect decisions that led to a tragic loss of human life and to a catastrophic financial loss. In still other cases, pure luck seems to have been guiding a jet to its safe landing.

Such seems to have been the case when a Cathay 727 slid into Hong Kong harbor in November 1967, to be followed a year later in November 1968, when a Japan Air Lines DC-8 approaching San Francisco airport landed on the bay instead and skidded to a safe stop. More than 200 passengers in these two landings escaped with no more than wet feet.

Yet in two over-the-water approaches within a few days of

each other in spring 1966, another 727 and another DC-8, coming into Tokyo's Haneda Airport didn't make it and in the fiery crashes that followed, 197 persons died.

In such tragedies, who is to blame? The pilot, for taking his plane and passengers into bad weather? The control tower, for not giving each flight its undivided attention? It is clear only that the crucial factor of human error must be given considerable more attention in the jet age.

Although all jet accidents are far from similar, there appears to be a pattern of misbehavior of the jets and the flying ability of the pilots during certain stages of the flight. It is unfortunate that in this modern age of computer technology a way has not been found that could instantly pinpoint an impending series of circumstances that could lead to trouble. For instance, in a survey of 153 jet airliner accidents between 1959 and 1965, only fifteen were in the cruising period. All the rest, 138 of them, were caused during takeoff, climbing, approaching, and landing. Over half of these mishaps occurred during the landing phase by undershooting the runway, running over the end, loss of control on the runway surface, or undercarriage problems. Almost every one of these accidents was due primarily to the failure of the crew at the controls, though poor or misread instrumentation and weather played a part.

The problem that can develop between the pilot and his jet may have had its beginnings during the transition stage between the pistons and the jets. It is significant that from 1959 until 1965 there were 144 jet accidents around the world and nineteen of them were during a pilot indoctrination period. Of the 265 crew members lost in the fifty jets that were destroyed in this group, thirty-two of them were pilots taking their transition training. More important, these pilots were lost while under the guidance and supervision of instructors who were the top performers in the business. It is clear, therefore, that under the most ideal of circumstances—good weather, new aircraft, top supervision—the pilots and their jets get into trouble and some particular brand of jet may be worse than another, trapping the unwary or the uninitiated.

Problems that may never occur in the transition training may show up in a scheduled flight.

When pilots are blamed for accidents—and they are held responsible for about 80 percent of them—extenuating circumstances seem to be present on most occasions. When you consider how many things can go wrong with a complex machine of this kind, the pilots are performing a fine job. An example of quick evaluation of a serious incident occurred when a Japan Air Lines DC-8 made an emergency landing at Oakland International Airport after the inboard right engine caught fire soon after a takeoff from San Francisco. The jet roared upward from the runway at 3,000 feet a minute, banking slightly to the left. Four minutes out, a fire was seen spewing from the engine as the aircraft reached a point directly over the San Bruno Mountains. Disintegration of the engine began to take place.

But the crew was unaware of the fire. The fire-warning system did not function. One or two passengers and the purser in the First Class lounge saw the flames pouring from the engine, and they, in turn, informed the flight crew. Quickly, the pilot made a left-hand turn to go back to the airport. Then, on second thought, he decided against another left turn. He thought that the still-smoldering engine might have caused some structural weakness. Instead he headed for Oakland Airport across the bay, landing to the sound of exploding tires, with full fuel load without thrust reversers. The passengers used escape chutes to reach the runway. No one was hurt. Pure luck.

The mishap was similar but perhaps not as frightening as that of the Pan American 707 that took off from the same airport on June 28, 1965. Fire immediately broke out in the number-four engine, the outboard engine on the right wing. Flames billowed upward and rearward with such intensity that the engine was burned away. Then twenty-five feet of the wing burned off and fell to the ground before the horror-filled eyes of 153 in the aircraft, sixteen First Class travelers, 127 tourists, and ten crew members. As happened in the DC-8 of

Japan Air Lines, the jet engine had come apart, but in this case it spewed white-hot parts into the fuel cells of the wing.

The fuel cells had been filled with kerosene in accordance with Pan Am's January 1965 edict that kerosene would be used instead of JP-4 "whenever possible." It is believed that the use of kerosene may have saved the plane from total destruction, as JP-4's burning rate is fantastic. Captain Charles H. Kimes, at the control of the 707, skillfully brought it to a landing at Travis Air Force Base, not wishing to make any turns with so much wing surface missing. That he was able to control the aircraft at all was attributed to a balance caused, on the one hand, by the loss of the heavy outboard engine, pylon, and wing, and, on the other, by the asymmetric power of the two engines on the left side and the remaining engine on the right side. Many of the lucky passengers, who expected to die, produced some spectacular movies and color photographs of the near disaster. The mishap proved again that crews flying modern jets are skilled men, able to think clearly in emergencies and act accordingly.

On November 9, 1963, a captain's skill was dramatically demonstrated on an Eastern Airlines DC-8 en route from New York to Mexico City. Captain Mel H. French, a veteran piston and jet pilot, was in command of the flight and he ordered an unscheduled stop at Houston to take on more fuel. His copilot was Grant Newby, and it was he who was handling the controls when the giant airliner, loaded with 125 passengers, went out of control.

According to the investigators, the jet took off from Houston and began to climb in a shallow ascent at approximately 700 feet per minute. The weather looked bad, even on the plane's radar. When the aircraft reached an altitude of 1,000 feet, its rate of ascent became even more shallow, climbing for the next minute (according to its flight recorder) at only 500 feet per minute.

Two minutes and twenty seconds after lift-off, the copilot pulled back on the control column, and the DC-8 began to climb through clouds at close to 1,000 feet per minute. It maintained this rate of climb until it reached an altitude of 4,800 feet, when it began to shake severely from bone-jarring turbulence. At this moment, Captain French was arguing with the Houston controller about his course, because he could see on the radar scope that his assigned heading was steering him into visible turbulence. The passengers were, of course, unaware of these arguments, though they are common occurrences.

The jet flew level for one second, then was put on climb again, but an even shallower climb than before, as more severe turbulence began to shake it like a cat worrying a mouse. The aircraft then began to receive severe convection drafts, which carried it up and down some two to three hundred feet at a time. Precisely eight minutes and forty seconds after takeoff, the aircraft received a punishing series of violent bumps, the severest striking when it had reached the altitude of 5,700 feet.

Captain French was concentrating his attention on the radarscope. Newby was gripping the controls. French's first intimation of dire trouble came when manuals, tools, maps, dust, engineer's books, and other articles began floating around in the cockpit.

When he looked at his airspeed instrument, it showed a "zero." Actually, investigators from Eastern and Douglas Aircraft later agreed the airspeed indicator was pegged at 400 knots—the highest it could go. The pointer was thus located at the top of the dial right beside the zero number. French saw the zero. He realized quickly that the plane was in a death dive. The altimeter was a revolving blur. French and Newby pulled back on the control column together but they were unable to budge it. The plane was now diving at close to six hundred knots, over the speed of sound.

French was a quick thinker. He only had a few thousand feet between his aircraft and the ground. He reached out with his right hand and reversed all four engines. This slowed the screaming jet slightly and permitted the nose of the aircraft to lift a bit as the two pilots pulled with all their strength on the control column. In a moment they were able to pull the yoke back, and the jet began to level out. They were now down to 2,300 feet above the ground.

As the jet was gradually pulled into its level attitude just over the trees below, the stresses upon its wing and engines became unbearable. One engine snapped off and plummeted to the ground. This was the inside right engine, next to the passenger compartment. Another engine was damaged.

One of the passengers said: "It felt as though someone had cut the elevator cable. When the jet dived, magazines, trays, pillows, books, small luggage flew all over the passenger compartment. Some of us who had failed to fasten our seat belts found ourselves glued to the ceiling. Some were actually in the aisles walking to their seats when the plunge happened, so sudden was the advent of the severe turbulence." (But what were they doing out of their seats so soon after takeoff in obvious bad weather?)

The plane, now level, was flown to Barksdale Air Force Base at Shreveport, Louisiana, where emergency equipment was more readily available at the moment. Seventeen passengers were injured; four needed hospitalization.

But what was important was the fact that the plane had behaved like the United jet over Nebraska and could be compared with the disastrous dive of the DC-8 outside of Montreal. Both United and Eastern pilots reported that they had lost their normal attitude references because the instruments acted crazily in the turbulence. And the big jets had very little safety margin when it came to abrupt changes in attitude. For instance, the normal descent of a jet airliner is between ten and thirty degrees downward from the artificial horizon (the level attitude of the plane in normal flight). The danger zone is reached when a jet begins to dive between thirty and sixty degrees. A pitch-over results, and the giant plane plunges into a fatal power dive.

Therefore, maintaining a level attitude or keeping the descent above thirty degrees is extremely difficult during violent turbulence. And then when the instruments begin to misbehave, it is often impossible for the crew to read the artificial horizon, and involuntary dives can actually be accentuated, unknowingly. Pulling a jet out of so violent a dive is difficult, even with thousands of feet between the plane and the ground. The Air Force found that their B-47s, without dive brakes, could pick up so much downward speed during involuntary dives that it took some four miles to bring them into a level attitude.

Sometimes, competent pilots cannot locate the trouble in time. For instance, on December 21, 1961, a British European Airways Comet 4B crashed at Esenboga Airport, on the outskirts of Ankara, Turkey. All seven crew members were killed, twenty passengers died in the flaming wreckage while seven were thrown free and survived.

The Ministry of Communications of the Turkish Republic immediately set up an investigation team, which was assisted by members of the Accidents Investigation Branch of the British Ministry of Aviation. The team probed the wreckage at the crash site and then removed all the parts that were required for detailed or laboratory examination.

The group found that the aircraft was properly loaded, and its center of gravity was correct to the limits set down in the manufacturer's manual. The gross takeoff weight of the Comet when it left London for Tel Aviv was below the maximum established for the flight, and no change occurred in this weight when it landed and took off from Rome, Athens, and Istanbul. Investigation further showed there were no defects in the aircraft and no repairs had been made to it during the routine flight. All flying surfaces appeared to be normal and undamaged prior to the violent impact. Painstaking examination of every part showed that the Comet was in excellent flying condition.

What caused the crash? Its commander, Captain K. J. Ruddlesdin, thirty-nine, had been with the airline since 1946 and had 785 hours in Comet jets. His copilot, Frank McTavish, had 961 hours on the Comets, and his other first officer, Colin Mervyn Bell, had amassed a creditable 536 hours. There was no doubt that the crew knew their flying. But did they know their jets?

The Turkish investigation team found, after assembling all

the available technical knowledge concerning the flight's last moments, that the disaster had been caused by the obstruction of the pitch pointer in the Captain's director-horizon instrument. This simple obstruction caused the Captain to climb too steeply after takeoff from the airport and, at the low altitude of 450 feet, the giant Comet stalled and fell in a perfect level attitude. The British, perplexed by the crash, found that the pitch pointer had been obstructed by one small screw, which had loosened. In fact, they determined that the tiny screw had not been tightened properly by the manufacturer. This infinitesimal object had caused the destruction of a proud jet and a heavy loss of life.

It was apparent, even back in 1961, that drastic changes were needed in instrumentation, changes that would increase the size of the artificial horizon and the directional and pitch-reference systems. There were some who thought that the artificial horizon instruments should be colored to show the sky and the earth and, in fact, some airlines did adopt this color-keyed system. But it was to take several years before American authorities got around to polling pilots on changes in artificial-horizon and other instrumentation in the jets. It was to take further disasters to awaken manufacturers, carriers, and government agencies to the need for information directly from the men who fly the jets.

Surely the lesson went back far enough. On March 2, 1953, a Comet 1A, with five Canadian Pacific Airlines crewmen aboard, along with five De Havilland crewmen and a flight engineer, crashed at Karachi, Pakistan, while on a delivery flight from England to Sydney, Australia. A long involved battle between the Canadian Air Line Pilots Association finally won from De Havilland the agreement that certain jet-flight techniques at takeoff had to be changed, and that the CPA crew was not to blame for the disaster.

There are instances where a mechanical failure in a jet takes place at the precise moment when a particular mechanism is urgently needed and the crew, unaware of what is happening, tries desperately to save the plane and its passengers. Such was the case of the Sabena 707 which crashed close to

Brussels airport on February 15, 1961. Sixty-one passengers and a crew of eleven died. A number of U.S. skaters enroute to the Olympics were on the flight. Efforts by the CAB to get an official cause of the crash were to no avail but it was believed that the failure of the air spoilers (air brakes on the wings) to work caused a loss of control during a turn into the runway on the final approach.

It is difficult to find the reasons beyond a shadow of a doubt when the pilots die as they did in this crash and as they did in the aborted takeoff of a 707 at Paris' Orly Field on June 3, 1962 when 122 passengers and a crew of nine died, including the cream of Atlanta's art colony. In this latter case, it is thought that the pilot tried to stop his takeover after he experienced heavy loads on his elevators but the decision was too late and the aircraft struck approach lights at the end of the runway and blew up.

There is a case on record of a pilot who aborted his takeoff and lived to tell his reasons. Although it was in his best judgment to terminate the flight before it roared any farther down the runway, he was nevertheless "pressured" into resigning at the peak of his career and after many years of superb service.

The facts are interesting. In November 1963, Captain "Mickey" Found of Trans-Canada Airlines aborted his takeoff at London Airport when his heavily loaded DC-8-F fanjet airliner "just didn't feel right." The plane failed to respond soon enough to its reverse power, ploughed through a field of brussels sprouts and ended up at the edge of a village. There was no one killed and no one hurt. Cost to repair the damaged airliner: $4,000,000.

After the investigation, Captain Found resigned. He left a post in which he had been flying for two decades during which time he had been one of the finest pilots in the business. But on this occasion, as his giant jet was hurtling down the runways with a full load of passengers, he pulled back on the column to lift it into the air and there was no response.

Captain Found decided not to play games at this juncture. The jet should have lifted and it had not; he was not going to count off any more seconds. He reversed the power and jammed on the brakes, but the terrific momentum of the DC-8-F carried it over the airport boundary.

Here was the case of a pilot who lived, who had honestly thought it was time for his plane to lift-off. It had not. He aborted the flight . . . and lost his high-paying career job.

Captain Found's case turns attention to the interpretation of instruments, the behavior and accuracy of instruments, and the human factor in any jet while under power. Human reactions will differ according to the circumstances of the moment. The USAF Medical Safety Division made a study of human reaction in flying techniques and arrived at some interesting conclusions. It takes one tenth of a second from the time the human eye sees an image to record it in the brain. It then requires a full second for the human brain to recognize the message, and five more seconds for the brain to signal the required muscles, and another four tenths of a second for the muscles to react to the message. Air reaction time, or the speed in which there is immediate aircraft response, is another five seconds. Total reaction time is 11.5 seconds.

It is apparent that the professional airline pilot has not much time to react to his instruments, make a decision, and translate that decision to the plane's controls. In that 11.5 seconds, the aircraft has moved a long way forward or a long way downward.

Yet, the ability of the pilot is no better than the messages he receives from the instruments. And often, when instruments and pilots are both at their capable best, tolerances for error in landings and takeoffs are so slim that crews must have their hearts in their mouths. It's no secret that a lot of commercial pilots take off with their fingers crossed.

One test pilot reported that a distraction of ten seconds could lead to a nose-down dive and the build-up of highly dangerous airspeeds. And events happen so suddenly that a pilot can become bewildered by the changes being performed by instruments and aircraft.

It has often been suspected that the pilots of modern airliners believe they are akin to gods—overly snobbish and aloof. The fact of the matter is they are human beings and are subject to stress and strain and home troubles and business problems and fatigue, just like any one else. Unfortunately, pilots have no time to daydream while controlling mighty jets.

That they are only too human is borne out by investigation of crashes over the past few years. One senior airlines captain was at the controls of an airliner that took seventy-nine persons to their deaths. It was discovered that up to the time of the disaster he was taking hypnotic treatments from a qualified doctor because he had become fearful of heights.

In another recent instance, the pilot of a plane that crashed and took a number of lives had serious marital troubles, was at odds with his wife, and was living with a stewardess who at the time of his crash and death was pregnant. Another senior captain on one of the nation's top flaglines was up to his ears in business ventures, and a few days before he died with two score others, he had made a very poor showing in his FAA examinations.

There is little doubt that the human factor is the important cog in safe air transportation. And the study of human fitness in relationship to the big jets is lagging behind, as the jets get bigger and faster and quieter. One might think that the faster and quieter the jets become, the easier on the flight crew. Unfortunately, the case is just the opposite. Cruising in the high thin air of the atmosphere in bright sunshine, with only a soft vibration and hum to disturb the quietness of the cockpit, can cause fatigue and nervous tension.

Dr. James E. Crane, medical examiner of the FAA, declares that preventive medicine is fine for airplanes, and the same principle is good for pilots also. He has reminded pilots that every component in an aircraft is replaced long before its failure can be expected and that this philosophy should also be applied to the human organism. Dr. Crane contends that there are many causes of fatigue in jets. Jets fly twice as fast as the piston-engine planes, doubling productivity and increasing the work load. Trips are shorter, and more trips are needed in order to fulfill the monthly hour quotas, the reason many old-line pilots would rather fly piston aircraft and get their required hours in with less work involved.

Then, too, Dr. Crane points out that east-west flights, crossing many time zones, raise havoc with body rhythm, sleep patterns, eating, and elimination schedules.

Pilots live in constant fear of infractions, he finds. They worry about outmoded traffic controls, noise abatement procedures, irregular trips, stand-by duty for days on end, refresher courses, and constant examinations.

Dr. Crane has since directed his probing pencil to the wives of commercial pilots, who have changed from piston to jet planes. What he finds is alarming. Significant changes in the health and welfare of their husbands is reported by 84 percent of the wives. Over half found their husbands more irritable; 67 percent of the pilots changed their sleeping habits; and 75 percent complained of being tired all the time. Wives further reported there was a changed attitude toward sex by one third of the group; 50 percent had altered their eating habits; 27 percent had changes in their emotional stability; and 39 percent did not enjoy their children as they had during the good old piston days.

Sitting in a 600 m.p.h. office watching and waiting was no picnic. Without a shadow of doubt, fatigue and worry were everpresent. It could build up, alarmingly in some pilots, to the point where their judgments could be warped. Before the Aerospace Medical Association, Captain Victor E. Schutz, Jr., said: "The care of such large numbers of flyers in their fifth decade of life poses a problem unmatched in the history of military and civil aviation."

Ten years ago, one pilot in every four was over forty. Today, half of all commercial pilots are over forty and many are in their fifties. Instead of progressing in life's work to a rather secure and mellowing stage, they have progressed into a rat race of lights and switches and noises and regulations thick enough to confound a mental wizard. The obvious solution would be to promote jet pilots to piston engines after they reach forty. But this might not be practical. Meanwhile, very little is being done about the problem of jet-age human factors. Yet the flow of information to the pilot, and from the pilot in the human-monitored system, is still the essential ingredient in safe performance.

The psychological and neurological aspects of the problem are complex enough when they embrace hearing, vision, balance, pain, postural awareness, time sense, and the barrier of inexpressibility. Added to this complex problem of the human factor is the control and chaos of air-traffic control with the everpresent chance of midair collisions, airport delays, all-weather landings, and modern airports themselves. Sometimes the aircraft controls do not react quickly enough to the pilot's touch to avoid trouble.

Shortly after 11:00 P.M. on the night of Tuesday, April 7, a Pan American 707 jetliner, delayed by bad weather, was finally cleared to land at Kennedy International Airport's Instrument Landing Approach runway under radar vectoring. The runway was called 4-Right. The jet appeared on the radar and seemed to be on course with the right altitude. Suddenly, the big Boeing vanished off the controller's radar because at that moment, with 145 persons aboard it was settling on the slick runway and rolling along at close to 150 miles an hour. It quickly ran off the 8,400 feet of runway, crossed a stretch of grass and 1,000 feet of swamp, coming to rest in a drainage ditch with all four engines coolingly immersed in water. There was no fire, but thirty-seven passengers were injured, some seriously.

On February 12, 1963, there occurred over the Florida Everglades a strange midair contest between a spanking new Boeing 720B jet operated by Northwest Airlines and the controllers at Miami International Airport with the weather acting out its part. The incident is a classic example of the confusion that surrounds the reading and understanding of radar, both in the aircraft cockpit and in the Air Traffic Control Centers, and of the irritating delays in getting flight corridors changed during an impending dangerous situation. It further points out that what one pilot thinks is good flying weather, another may think is anything but.

There will always be arguments concerning the wisdom of

Captain Roy Almquist in taking off from Miami International Airport in such beastly weather. The weatherman had been predicting it would be severe and, as far as the human eye could see, the sky was black and foreboding. Flights after this one were diverted because of the same storm, and at the very moment that Almquist was getting ready for his takeoff, an Eastern Airlines captain seated in his 707 jet on the same airport could see a wall of water in his radar and decided to wait on the ground rather than take a chance on the storm.

But after talking with the forecasters and studying the general weather patterns, Almquist elected to operate the flight to Chicago with the reservation that he could take off into the west and make an immediate left turn to fly southwest, thus circling around what he thought would be the worst area.

Sharp at 1:23 P.M., the flight called Ground Control at the Miami Tower, just as Almquist was about to start the engines. Copilot Robert Feller asked how they were routing outgoing aircraft at the moment, as he was going to Chicago and would like to be clear of the storm that was so obviously dominating the skies to the west and northwest. The Ground Control reported there was a pretty thick line of storms to the northwest and as a result departing aircraft were being shuttled to the southeast or the southwest and then climbing back over the top of the storms.

"That's what we had planned on," Feller replied, and he was told to call the controller again after the flight had departed from the terminal and was on the way to the takeoff pad. Almquist and Feller kindled the mighty power plants and the jet was pushed back from the terminal sufficiently far for it to maneuver under its own power. As it moved ponderously from the terminal buildings, the Ground Control reported that the flight's takeoff runway would be Twenty-Seven Left into a twenty-five-knot wind coming from a compass heading of 290 degrees.

The flight switched from Ground Control to Departure Clearance Control, which would be in charge of the outbound flight until it was well beyond the control of Miami Airport personnel. In contact with Departure, the Northwest flight was informed that it was cleared to Chicago's O'Hare Field over various routings that were now being spelled out, routings that would send the plane up over the Everglades, northwest to St. Petersburg, and on to Chicago.

The flight was advised that after takeoff it was to turn right and maintain 3,000 feet of altitude until told to do otherwise. Feller and Almquist evidently did not like this routing one bit. A right turn would head them directly into the blackness of the storm, and besides they had only just been informed that all flights were taking off southeast or southwest. Why the switch?

Just before taxiing onto the waiting pad at the east end of the active runway, Feller called the tower and pointedly asked for a routing to the southeast, a more comfortable distance from the visible storm path. This was common sense, and the tower asked the flight to stand by while the tower ascertained if there was an air corridor open to them. "We'd rather hold than make a right turn out," Feller radioed the tower.

The tower replied in several seconds that the flight could, indeed, make a lefthand turn and head due south after takeoff until notified of the new course being prepared for it. Almquist shoved all four throttles full forward, and the 720B screamed down the runway into the west wind and lifted easily at the first intersection. It climbed over the airport boundary, over the Palmetto Expressway, and then made a lazy turn southward and just a little southeastward, maintaining the low altitude of 3,000 feet. A few seconds later, the controller straightened the course to a due south heading over Homestead, Florida, and gave the aircraft permission to climb to 5,000 feet. The nose went up and the jet climbed to its new assigned height, the storm building mightily on its right side. The flight then took a turn to the south-southwest on orders from the tower. In a few minutes, it was ordered into a turn to the west.

"If we could go up now, we'd be in good shape," radioed Feller.

"Okay, stand by," replied the tower.

At this point, the tower became occupied for a moment with another flight, a small aircraft heading out of Miami on a north-northwest course at low altitude, and while this conversation was in progress, the Northwest crew noticed that the heaviest blackness now appeared to be directly in front of them.

"It looks like we're going to run back into this at this altitude, is there a chance to go back southwest or east?" Feller pestered the controller.

The tower attempted to be reassuring and told the Northwest crew that they would enter an area of rainfall four miles ahead on their present course, then be in the clear for about three miles, and then back in heavy rain again. The flight should break out of this patch in a matter of minutes and be clear for the rest of the trip.

But Feller and Almquist did not like the blackness of the storm, which meant they didn't believe the radar. They would be flying level at 5,000 feet near the blackness of the swirling clouds, through tremendous castles of gray and blue and black.

"We're in the clear now," radioed Feller, "we can see it out ahead . . . it looks pretty black."

"O.K. Northwest Seven Zero Five, we're working on a higher altitude now," the tower called through the static.

In a matter of seconds, the flight was advised to start climbing again, with a turn to the left to avoid the black patch that Feller had reported directly in front of him. This decision was unfortunate, because the flight ran smack into heavy turbulence and reported it to Miami Center.

The controller suggested an immediate right turn to get out of the trouble, but the turbulence was now so severe it was considered too dangerous to make any turn that courted the possibility of losing control of the aircraft. A severe updraft or a downdraft during such a turn could flip a jet on its back in the twinkling of an eye, and Almquist and Feller were not going to take this chance.

The controller was sympathetic, and said that he had vectored the jet into the area of least turbulence on his radarscope, an indication of how bad it was in the rest of the area.

"Okay," said Feller. "But you better run the rest of them off the other way then."

When Northwest's flight reached 10,000 feet, it was forced by rules to switch from Departure Control to Air Route Traffic Control for the remainder of the flight out of the state. The flight was advised to make a turn to the north by northwest and did so, beginning its turn after passing through the 16,000-foot level. The time: 1:48 P.M.

There was trouble in raising Air Route Traffic Control because of bad static, and by the time that a clear frequency was established between the Center and the flight, the jet was climbing through 17,500 feet. It had just reported this fact, when all of a sudden, the air was filled with a jumble of unintelligible words.

They sounded like "tank two." Was it "thank you"? Or was there another flight butting in on this frequency? In any event, the Air Route Traffic Control wanted to get the message clear and called the flight. "Northwest Seven Oh Five . . . this is Miami . . . Northwest Seven Oh Five, Miami . . . Northwest Seven Oh Five, Miami . . . North . . ."

There was no answer from Northwest 705. At this moment, Almquist and Feller were fighting to save their big jet. They were in the grip of violent updrafts and downdrafts at an altitude of 19,500 feet. They were trying to bring the bucking monster under control, while the rest of the crew and the thirty-five terrified passengers were holding on for their safety and their lives.

The next episode in this tragic drama occurred on the ground or—to be more exact—from the deck of a cabin cruiser on the Shark River, in an area of the Everglades known as The Banana Patch, located some fifty air miles southwest of Miami.

Three couples on the cruiser were trying their luck at fishing. The movement of the thunderstorm front from the northwest caught their attention and, as it moved closer and closer, there were thoughts of increasing speed and heading out of the swamps for Whitewater Bay on the southwest coast of

Florida. From there, the group could head north and back to their cottage in Naples, where they were vacationing.

Someone shouted, and all eyes turned to the mass of black thunderclouds above. A ball of fire had appeared. It was falling lazily toward the ground. Someone thought the Cubans were attacking. But no one laughed. They just stared at the strange object.

The flaming object struck the ground about a dozen miles away, and a tremendous explosion reached the ears of the fishing party a few seconds later. They realized that the falling object was an airplane and that the proper thing to do at the moment was to get a "fix" on the area where the craft had gone down and then report the incident to the authorities.

What had turned the attention of these people to that part of the sky where the ball of fire appeared out of the black clouds? They agreed, in talking it over, that it was an explosion in the sky that had first attracted their attention. This was important, because an explosion in the sky indicated the aircraft had broken up in its descent and that its fuel cells had blown up.

The group headed back down the Shark River, crossed Whitewater Bay and traveled through lagoons to the United States Ranger Station at Flamingo where they turned in the alarm. The big jet had been missing since approximately 1:55 P.M. It took Miami Airport authorities an incredible sixty-six minutes after the jet had failed to respond to radio signals and failed to appear on their radar sets to take action. Some of this time delay could understandably be excused by the fact that dense areas of precipitation cut off the radar scanning. Also heavy static had been interfering with radio transmission.

The United States Coast Guard had two planes in the air at the time, both of them amphibians. One was cruising in the St. Petersburg area and the other off Miami. At 2:48 P.M., when they received news that the jet was missing, these two aircraft were diverted to the Everglades to search. It was not until search authorities received the "fix" from the fishing party at Flamingo that they had a fair idea of where the plane had fallen. Then, as darkness and heavy rain closed in on the lonely region, a Coast Guard helicopter, making its last run over the sawgrass ten miles north of Shark River, spotted smoldering bonfires. A crisp report was radioed to an airport at Forty-Mile Bend on the Tamiami Trail, where search headquarters were being hastily set up. It said: "Found plane . . . no survivors."

10

The Death of Flight 304

Jet airliners are designed with thin swept-back wings and long lean passenger compartments to slice through the thin upper atmosphere at speed close to the sound barrier. Because of their aerodynamic shapes, they are subject to a phenomenon that quickly removes their ability to maintain lift.

The phenomenon is a stall, a word that is impressed on every airplane pilot, from the day he starts to learn how to fly. In conventional aircraft, the stall occurs when the machine is slowed down to a speed that is below the minimum to maintain flight. The aircraft will fall to the ground if the speed is not immediately boosted either by an increase in engine power or by a dive to get the wind forces working efficiently over the wings again. Jets will stall at low speed also, but unlike conventional aircraft they will also stall at very high speeds.

It seems incredible that a modern jet airliner can stall while traveling at high speed but the answer lies in its arrow shape which is designed to fly at precise speeds at altitudes between 26,000 feet and approximately 41,000 feet. What keeps the aircraft from stalling is a combination of power and attitude, the latter being the precise horizontal position of the aircraft to part the barrier of air with as little effort as possible.

As every pilot knows, raising the nose of the aircraft upsets the optimum horizontal position and in this position speed can fall off rapidly unless more power is added. But in a jet the addition of power in very thin air can cause the air to flow across the wings and bleed away from the wing tips destroying the supporting cushion. Therefore a pilot would not wittingly raise the nose of his jet when he is flying at the maximum speed range for a given altitude. But a deadly trap lies in store for the pilot who finds his aircraft caught in turbulence and being forced nose-high. Now he is in a spot. Jets can't be swung sideways and dived like a single-engine biplane. Great care and control must be exercised at all times by the crew, whether the aircraft is on automatic pilot or not, because jets have a built-in horizontal stabilizer that operates instantly when it senses that the aircraft needs control over its tendency to nose up or nose down. This device has been mentioned previously. It is the pitch trim compensator. However, many pilots would rather depend upon their own sensing systems than that of a machine.

Low speed stalls and high speed stalls are more likely to occur between 30,000 feet and above. The denser the atmosphere the less likely it is that this condition can take place. Therefore, the higher the jet flies toward its service ceiling, which is 43,000 feet, the more likely it can stall if the pilots jockey their speed and attitude during turbulence.

For instance, a giant 270,000-pound jet airliner, typical of today's fleets, flying at an altitude of 35,000 feet above sea level, is encased in a one-thousand-foot corridor where it must fly at more than 185 knots an hour so that it will not suffer a low speed stall and it must not fly more than 295 knots or it will succumb to a high speed stall. It is therefore flying in a stall-free range of 110 knots—not much if the aircraft is subjected to speed variations due to turbulence.

If the jet should be at a much lower altitude, say at 31,000 feet, the stall-free corridor widens to a spread of 161 knots. But if it should be flying at 39,000 feet which is not unusual for today's jets, and flying at its recommended cruising speed of 560 m.p.h. (knots have been translated to miles per hour here to better explain the phenomenon) the stall-free zone is a scant seventy miles an hour. In other words if the jet slows down to approximately 530 miles an hour it could suffer a low speed stall. Conversely if it increases its speed up to 600 miles an hour it could be trapped in a high speed stall.

Pilots become aware of the stall by the onset of a stall buffet, shuddering that envelops the aircraft split seconds before the actual stall takes place. Delicate instruments also sense the stall and warn the flight deck occupants with a raucous blowing of a horn.

It may seem to the reader that a jet airliner is flying in a corridor that is much too narrow for safety, leaving little leeway in case of an emergency, but the plain truth is that jets are designed to fly under such conditions because large variations in their cruising speeds are most unlikely.

Another factor rears its unwelcome head at this time. We have seen that altitude, speed and horizontal attitude affect the stall boundaries but the weight of the aircraft can also affect the relationship. We know that weight cannot be added to a jet while it is in flight but "apparent weight" can be added by the pilot pulling back on the control column, which raises the nose like a vertical wind gust could do. The possible increase in "apparent weight" can drastically narrow the stall corridors.

Not only that, but various maneuvers of a jet can set up loads that are similar to adding weight. Banking at forty-eight degrees to the left or right, perhaps to avoid a possible mid-air collision, could immediately set up forces that change the weight balance of the aircraft and cause a loss in required speed. A slighter turn of twenty-five degrees made at the same time as a gust of wind catches the wings would create the same degree of danger as the first example. This is why pilots are warned not to bank their aircraft in turbulence or gusty wind conditions.

It is clear therefore that jets have certain design limitations and that when they are exceeded, disaster may follow. Jets were built to fly in good weather. When turbulence occurs, the recommended speed penetration today has been increased from 145 knots to 280 knots (380 miles an hour approximately at jet cruising altitudes) up to 35,000 feet to overcome low speed stall possibilities. This speed is rough on the passengers but the aircraft has a better chance of survival. It is better to take a chance on the structure holding up, than to risk a dangerous stall situation.

To maintain his straight and level flight through the air, the pilot relies on an aerodynamic feature which is built into every aircraft. It is known as trimming, the art of altering the attitude of the aircraft to its longitudinal or horizontal plane by raising or lowering the nose to a position where the wings best attack the air mass through which they are moving. This trimmed adjustment is called the angle of attack. The raising or lowering of the nose is accomplished by a stabilizer trim wheel at the pilot's side which activates an air foil on the tailpiece.

When an up-current of air raises the nose, the pilot moves the trim wheel forward and thereby returns the aircraft to an even keel. When the nose dips in a downdraft he moves the wheel rearward to regain the best attitude. In a jet this maneuvering under turbulent conditions can be a deadly trap.

In the good old days of the propeller-driven aircraft, constant trimming during the entire flight was commonplace and safe, primarily due to the sluggish response to controls and the fact that the aircraft had some gliding characteristics that jets have never had. But jet pilots are graduates of the propeller era where constant trimming was a way of life. This has led experts to believe that many recent jet disasters may have been caused or accentuated by trimming during critical speed periods near the stall-buffet zones. Mysterious jet dives had all the earmarks of overtrimming and the FAA issued a set of turbulence penetration instructions in the hope of overcoming long years of flying procedures.

"Flying under turbulence conditions," states the FAA, "requires techniques which may be contrary to the pilots' natural reactions. The natural stability of the airplane will work in a direction to minimize the loads imposed by turbulence. The pilots should rely to a major extent on this natural stability and not be too greatly concerned about the attitude variations. Since there is always uncertainty of the direction, timing and size of the next gust, it is often better to do nothing at all than to attempt to control airplane-attitude too rigidly."

The FAA recommends that attitude should be controlled in turbulence with the elevator and NEVER with the stabilizer trim wheel because any updraft or downdraft that would tempt the pilots to change the trim could be expected to reverse itself in the next few seconds. If trimming counters the first draft, the second, which is always opposite in direction, will exaggerate the trim condition and the jet will be pointed too high or too low.

Nose too high means a stall; nose too low means a death dive.

The problem of trimming, or rather the controversy concerning trimming, never became so meaningful until the public inquiry into the plunge of an Eastern Airlines DC-8 into Lake Pontchartrain shortly after takeoff from New Orleans during the early morning hours of February 18, 1964. Forty-nine passengers and seven crew members died. The Flight Recorder was destroyed; therefore, the behavior of DC-8s actuated by trimming became the subject of a long and involved debate between pilots, airlines, government agencies and aircraft manufacturers.

The flight, operating as 304 from Mexico City to New York City, had landed at New Orleans after midnight and was ready to continue the journey at 2:00 A.M. The tower was informed that the four engines were being started. A flight plan had already been given to the pilots routing them out of New Orleans on a course to Atlanta and thence to New York. Captain William B. Zeng, forty-seven, of Ringoes, New Jersey, was in charge of the flight. At his side was First Officer Grant Newby of New York City and behind him was pilot-engineer Harry Idel, of Farmingdale, Long Island. The tower reported that the runway would be Number One, indicating a takeoff into the north, and the winds were blowing between ten and fifteen knots from the northeast. There was a solid cloud deck between 1,000 and 6,000 feet with turbulence associated with the mass. Rain showers had been reported in the lake area. Otherwise, conditions appeared normal for flight operations. The flight took off over Lake Pontchartrain and in about thirty seconds disappeared in the clouds. Radar of Departure Control took over the handling of the flight. The New Orleans Route Traffic Control Center also watched this flight as well as another in the vicinity. At that time of the morning there are not many flights to watch. After a Delta jet was given its heading for Houston, the Center called Departure Control asking if the Eastern flight was on a specified heading out of the area.

TOWER (*Departure Radar*): Yeah, zero three zero.

CENTER: Just leave him on that heading.

TOWER: All right. Eastern Three Oh Four . . . continue heading zero three zero for Center vector and contact New Orleans Center radar frequency one two three point six . . . now.

EASTERN: O.K.

(*This was the last message from the flight. It was Captain Zeng's voice.*)

CENTER: Did you send Three Oh Four over? . . . He's not talking to me.

TOWER: Yeah I did . . . I'll shake him up again. . . . Eastern Three Oh Four, contact New Orleans Center radar frequency one two three point six, now please . . . [repeat] Eastern Three Oh Four, contact New Orleans Center radar frequency one two three point six. Give him a call Control, again please. . . . Hey Control you got him?

CENTER: No, I'm not talking to him . . . he disappeared off the scope . . . up there, northeast.

TOWER: Mine, too.

CENTER: I don't know what happened.

TOWER: Eastern Three Zero Four . . . one two three four five four three two one . . . Eastern Three Oh Four . . . New Orleans Departure Control calling . . .

TOWER (*to Local Control*): Give me a number for Eastern Air Lines. . . . Hey, Center, call the company will you, we're trying to get a phone number to call them now.

CENTER: We'll call them now.

TOWER (*to Local Control*): Hey Sig . . . Hey Siggy . . . are you painting him at all yet, control?

CENTER: No, I haven't painted him since about ten or twelve miles northeast.

TOWER: Yeah, the same here. . . . I saw him . . . I saw him fade but I didn't know whether he was out of my coverage or what.

CENTER: No, I even lost him on the beacon. . . . I had his beacon and I lost that too . . . so I don't know what happened . . . we got the company on it now and we'll keep you advised.

TOWER: Yeah, do that . . . please . . . Eastern Three Zero Four one two three four this is New Orleans Departure Control.

CENTER: Eastern Three Zero Four this is New Orleans Center. . . . On Guard if you hear . . . change to one two three point six . . . one two three point six . . .

TOWER: Hey Center, did you get anything at all?

CENTER: No we haven't got anything yet Approach . . . we've done all we could though.

TOWER: Yeah, okay. Do you want us to go ahead and call the State Police for the Causeway and have them run the Causeway, because he was in the vicinity of the Causeway on my scope . . . just east of it . . . yes he . . . he was . . .

CENTER: Pretty close over the Causeway, you might do that, yes.

TOWER: Yeah, alright, we'll do that now . . . we're starting procedures now on him . . . uh, everybody is unable to raise him and we're alerting the Coast Guard and various agencies for him.

CENTER: Okay, thank you.

This ended the conversations at the New Orleans Traffic Control Center and the Control Tower in connection with the strange disappearance of the Eastern Flight. The talking back and forth and the calling into the blackness of the night are only so many words on paper, but in reality the controllers on duty that night were a group of badly frightened young men peering through the tower windows or into the pulsing circles of their radarscopes, wondering what had become of a modern airliner that just simply vanished.

The first item of business now was to find the missing jet. According to the plane's last noted position on radar, it had disappeared somewhere over the Lake Pontchartrain Causeway, about twenty miles northeast of the airport. A police chief, Alfred Knight of the town of Mandeville, ten miles north of the Causeway's terminus, began to get calls reporting an explosion. But there was no report of fire. No report of wreckage.

At dawn, Coast Guard helicopters and several surface craft found a patch of water that appeared to be littered with debris. A man's sports jacket was floating beside a pair of trousers. There was a white shirt and some bits of paper. Oil coated the surface of the water. Using fishnets, the searchers dragged the shallow bottom of the lake, about fifteen feet at this point, and discovered a section of a human body. The crash site had been found.

On Tuesday, July 14, 1964, in New Orleans, the Civil Aeronautics Board held a public inquiry into the Eastern DC-8 disaster. Robert T. Murphy, vice-chairman of the CAB, conducted the hearings. In his opening statement, he read into the record that "The electrical system, radio communications, and navigation equipment, the instrument system, the oxygen system, and fire protection system all sustained considerable damage in the impact. However, nothing was found to indicate pre-impact failure or malfunction of these systems.

"All four power plants were recovered from the lake bottom with no evidence of inflight fire, damage, or other failure prior to impact. Examination of fuel tanks and fuel system showed no indication of burning, sooting, or arcing.

"Its hydraulic system showed no evidence of operational distress. Evidence indicates that the flaps were in the up position (normal flight position). Nearly all of the horizontal stabilizer-control system mechanism located in the tail section was recovered.

"This aircraft was involved in a reported stabilizer malfunction incident in San Juan, Puerto Rico, September 11, 1963, and shortly thereafter repairs were completed and the aircraft returned to service.

"The maintenance record revealed that the pitch-trim computer had numerous write-ups by flight crews for unwanted extension with the control switch in the normal position. A ground test was made in Philadelphia, February 24, at which time a mechanic found the system failed to check. An entry was made in the log book to this effect, and the aircraft was dispatched to Mexico City."

The first witness to testify after the conclusion of the Murphy statement was Eastern Airlines pilot Paul M. McGill. He had flown the DC-8 to Mexico City and handed it over to Captain Zeng and copilot Grant Newby. (Poor Newby—he had survived the plunge of the Eastern DC-8 over Houston and, in fact, had shown fine skill in bringing it under control. Fate had his number.)

Captain McGill confirmed that the jet's pitch-trim compensator was inoperative, but, he said, he had absolutely no qualms about flying the DC-8 with it not working. "The aircraft was normal all the way from Philadelphia to Mexico City," he said.

Chairman Murphy asked the pilot if he felt that the aircraft trimmed normally, and McGill answered that it did.

"Then could we say that the pitch-trim compensator is not a 'go' item?"

"That's correct," replied McGill.

Another Eastern pilot was called to the stand. He was Captain Louis Fedvary. He had flown this ill-fated DC-8 before it was routed to Mexico City, and on that flight his copilot had been Grant Newby. He was questioned about Newby's handling of a jet in turbulence, and the captain replied crisply:

"Newby flew a plane the way it was supposed to be flown."

Norman Smith, a pilot-engineer with Eastern, was questioned at some length by William L. Lamb, chief investigator for the CAB. Again the question was asked if the pitch-trim compensator was inoperative on the Mexico flight, and Smith replied, "Yes."

"Did you discover it?" asked Lamb.

"Yes, on the preflight check-out."

He further noted that the mechanic's log showed it was inoperative.

Under extensive probing by Lamb on the various engineering procedures that are used to check out a jet before takeoff, Smith related the exhaustive checks used by mechanics and by engineers to ensure the working or nonworking of all aircraft essentials. On this occasion, when he discovered the pitch-trim was out of order, he deactivated the system by pulling three circuit breakers. A teletype message was at once sent to Mexico City that the pitch-trim compensator was out.

He was then asked if there was any restriction placed on the jet as a result, and Smith replied there was a speed limitation. He was further asked by Lamb if planes of this type had chronic trouble with gauges, and Smith answered that the instruments were in frequent disagreement but ducked the use of the word "chronic."

Captain A. V. Appelget, supervisor of flying for Eastern, told the probe that even if the pitch-trim compensator burned out in flight, it would not aerodynamically affect the plane, and the pilot would be able to exert "more than adequate control." Appelget also revealed under questioning that it was common practice to place the DC-8 on "automatic pilot" during climbing periods.

A Northwest Airlines official whetted the interest of the listeners when he testified that commercial jet instruments give pilots wrong information under certain flying conditions. Paul A. Sonderlind, manager of research and development for Northwest, at Minneapolis, revealed that studies of air accidents had shown that jet instruments have registered incorrectly when planes were climbing. He said he could not confirm that the climb situation could cause the erratic behavior in the instruments, but he indicated that it might be a factor.

He said he had conducted a study of jetliner accidents over the past eighteen months and found that all the accidents tended to occur while the aircraft was in a climb, and, as far as could be determined, they happened when pilots were flying on instruments rather than using visual references. He told the board that pitch-trim compensators were necessary to the jets, and their use contributed to the stability of the aircraft. He said his studies also indicated that jet accidents could be the result of aircraft being pitched up or down by turbulence or some other factor. He flatly declared that a jetliner flying on instruments was not as likely to recover from a crisis as one whose pilots were using visual guides. His testimony raised the eyebrows of many who thought that the human factor should be more and more augmented by instrumentation.

Charles A. Ruby, President of the powerful Air Lines Pilots Association, snapped that most pilots would rather have pitch-trim compensators removed from their jets. Even if the pitch-trim compensators were improved, he told the board, he still believed that pilots would rather do without them. "The important thing to remember," he said, "is to keep complications out of the primary-control system. I do not know of any instance where the pitch-control compensator could be solely blamed for an unusual climb or the dive of an aircraft. My personal view is that to fly this airplane properly, you should be able to give it your full attention, especially at high altitudes."

Dealing with the matter of flying a jet in turbulence, Ruby said: "Our ability to read and interpret instruments in jet aircraft leaves quite a bit to be desired." He said it was his own experience that, under certain conditions, instrument panels could be a blur for as long as five or ten seconds during turbulence penetration. However, he called the DC-8 a satisfactory aircraft and confided he would be more comfortable flying it manually than having a lot of "little gismos" helping him.

The inquiry was enlivened by a charge against six U.S. flaglines by the maker of one of the pitch-trim computers that the service reliability of the compensator had been prejudiced by inadequate overhaul and test procedures. Arthur N. King, manager of the firm that manufactures such instruments for DC-8s, declared to the board that he was critical of maintenance procedures by airlines. He said their maintenance checks were not thorough enough to keep the equipment in the best working order. He charged that such trim instruments were not placed in the repair shop until defective, and that they should have routine maintenance on them long before they broke down.

John H. Armstrong of the Douglas Aircraft Company replied: "The aircraft has been demonstrated to be safe no matter what the actuator does."

Another witness, Arnold G. Heimerdinger, chief pilot of Douglas, declared he would object to the removal of the pitch-trim compensator in the DC-8s. He pointed out that precise attitude was difficult to achieve when the instrument was not in use. Along with other Douglas engineers, he explained the necessity for such a device.

Apparently, when a high-performance aircraft approaches the speed of sound, he said, the center of aerodynamic "lift" moves toward the rear of the airplane, and this behavior creates a "tuck-under" tendency. The pitch compensator changes the pitch of the aircraft to compensate for this change by sensing the change and trimming the aircraft to maintain its level flight attitude. This is constantly changing during the flight through high altitudes.

The mechanical part of this system consists of a simple rod, which pressures a spring against the steering column, to keep the nose of the aircraft up. All this is actuated by a computer, but if it is shut down or fails to work, the pilot can do the same job manually.

Charles Ruby, during his lengthy testimony, mentioned that he had received oral and written reports from pilots who had experienced difficulties with the device under certain flying conditions. These occurred when the nose of the airplane was forced up, when in reality it should not have been moved at all. One witness reported that, since June 1964, National Airlines had come to recognize inherent problems with the compensator.

Witness Orville Dunn, assistant chief of aerodynamics for Douglas, testified that the decision to use the spring-forced system in the pitch compensator was basically his. A graduate of the Massachusetts Institute of Technology, Dunn said there was nothing safer than a spring. He considered the compensator to be a safe operation, but he reminded the board of inquiry that a jet passenger plane was not a fighter aircraft and that its response was more sluggish. He said that the jet is more stable when the actuator was working than when it is inoperative.

And now the bombshell was to be dropped by Roy Peterson, the chief of the Flight Test Branch of the FAA West Region. He bluntly told the inquiry that the DC-8 stability was poor during a fast climb, but that this was not learned until after the jets were well along in commercial use.

Under examination by Martin V. Clarke, assistant chief of engineering for the CAB's Bureau of Safety, he was asked at the outset if any special conditions were placed on the DC-8 when it was certified for commercial use.

Peterson replied there were none, but there had been some amendments on technical points. He said the jet complied with all the special requirements of the Civil Air Regulations. He told the board that the FAA had conducted tests on the pitch-trim compensator in a "runaway" condition and found that the pilot would have to compensate for the device's malfunction manually within three seconds, or the aircraft would lose altitude or normal equilibrium. A runaway condition, he explained, was when the automatic pitch-trim device takes corrective action when no such action is needed.

Peterson was then questioned about the control of the jet at lower altitudes with the mechanism inoperative, and he replied: "It has weak stability in a climb at 300 knots [about 375 miles an hour]."

He said that the DC-8 demonstrates positive stability while cruising, but not during a high-speed climb, a condition that occurs a few minutes after takeoff. He was questioned closely about this behavior and was asked whether such tests were included in the flight testing when the DC-8 was certified for commercial use. "We specified the climb condition as stable at the time of the certification, as we felt there was no problem in the high-speed area, and it was checked at certification. We had no reason to believe that the stability would be weak in a climb at 300 knots."

Reid C. Tait, of the FAA, asked Peterson if the Civil Air Regulations bar planes from climbing at high speed, and the answer was negative. It was then revealed that the FAA had been conducting an engineering study of the DC-8 in cooperation with engineers of the Douglas Aircraft Company.

W. A. Bryde, aerospace engineer with the FAA at Los Angeles, said that he headed a committee to investigate all known aspects of the DC-8 crashes at Pontchartrain, Montreal, and Lisbon. The purpose of his study was to determine whether there were any mechanical characteristics built into the DC-8 that could have contributed to the crashes. The investigation was still far from complete.

The public inquiry into the Lake Pontchartrain disaster ended with a wide division between witnesses as to the necessity of the pitch-trim compensator which is still being used today on most jet airliners.

However, two years of intense investigation and many jet test flights proved that the inquiry had been following the right track. The CAB determined that pitch-trim had been somehow extended—but not by the pilots—and that when the aircraft entered the short sharp turbulence a few moments after entering the clouds, the jet nosed over and dived. Immediately the Douglas Aircraft Co. was ordered to change the limit of the pitch-trim from two degrees to one degree.

One degree meant the difference between life and death.

11

Air Traffic—Control or Chaos

What does a parachute drop into Lake Erie and the deaths of eighteen men have to do with Air Traffic Control?

Plenty, as we shall see.

Many pilots refer to Air Traffic Control as Air Traffic Chaos. If it is bad today, it will be worse tomorrow. So much so that industry leaders have predicted a complete breakdown in supervision of intercity air traffic by the early 1970s. It is a problem that can be solved only by education and an intelligent approach by the government, the airlines, and the public. A series of predictable disasters unfortunately may be the catalyst which will help solve the enigma.

Air Traffic Control is the vast and complex system operated by the Federal Aviation Administration to keep control over the traffic created by more than 130,000 airplanes and the half a million pilots who fly them. Only the fact that all are not in the air at the same time keeps the system from failing. Yet, on any busy day, an estimated 30,000 flights are conducted in the air space, under surveillance, known as Instrument Flight Rules (IFR). More than 5,000 of these are in the crowded air corridors of Boston, New York, Washington, and Philadelphia.

The public is only dimly aware of the overwhelming problems of trying to keep airlines on time and yet from colliding with one another. People realize from time to time that something is amiss when airport landing and takeoff delays of scheduled flights run anywhere from one to three to four hours even in good weather. The periods of time are longer during severe bad weather, with thousands of planes stacked up in holding areas all over the country in a complex mess. Actually, more than six hundred new airports are needed to siphon off the overflow. In a single hour, a harried Air Route Traffic Controller handling approach over a complex like Chicago will be in charge of guiding forty airliners and responsible for the lives of 4,000 people. At Chicago's O'Hare airport, the largest in the world, controllers handle 657,000 takeoffs and landings each year. The job is getting bigger month by month and the planes are getting bigger day by day.

In 1965, hundreds of dedicated young men handled 37,870,535 operations from the control towers, of which some eight million landings were instrument operations. By the end of 1967, the towers would handle almost fifty million operations with 11,661,002 flights under instruments.

Controllers realize that big jets and baby puddle-jumpers have the same rights on the airports, and they must guide great and small—with their tremendous differences in speeds and flying characteristics—in and out of airports squeezed by encroaching subdivisions and high-rise apartments. Controllers must depend entirely on radar, which is not infallible, and on good luck. All too often luck fails them. Every man in the system worries whether he will be the next one to cause a midair collision or a dangerous "near miss" of which there are now over a thousand a year—"misses" where evasive action had to be taken by one or both crews to avoid a midair collision.

One would think that the airlines themselves would be fighting for safer corridors by scheduling more flights during the dark hours and reducing the peak daytime congestion. Yet no move has been made to stretch out the schedule into a twenty-four-hour system, and the chances are that nothing will be done unless the government forces such a move. Not only that, but on the Atlantic where a 120-mile corridor is maintained for each transatlantic flight, airlines are battling for narrower corridors so that they can place more flights in operation. Transatlantic airline pilots claim such a move would cause the immediate deterioration of safe flying practices. The pilots have been battling for years against the reduction in width of air corridors . . . but they are doomed to failure.

With the exception of the United States, almost all the world's airlines are owned and operated by the governments of their countries, for profit and balance of payments, and no pilots' association or groups of them can beat down all the governments.

Just how chaotic is Air Traffic Control can be shown in the case of the Ohio parachutists on the afternoon of August 27, 1967.

Thick and heavy cloud layers hung over northern Ohio that afternoon, with clouds that billowed upward from about 3,500 feet to as high as 9,000 feet. To the twenty experienced parachutists who elected to fly up to 20,000 feet and beyond for a mass descent to Ortner Airport, near Cleveland, the weather must have looked good enough for jumping. Though it is illegal for chutists to drop into clouds unless in an emergency, apparently all of the men forgot the regulations. Of course, they were under government surveillance and no one seemed to offer any complaints.

Pilot Robert Karns, owner of a North American B-25, agreed to fly the group for the high-altitude jump, a job he had successfully done before. The drop had been set for Saturday, August 26, but it was rescheduled for the next day to accommodate the Ohio-wide influx of enthusiasts. Twenty-three chutists showed up. Karns, however, reluctantly told three of them they would have to remain on the ground as there was room for only twenty on his B-25. Those three who were left behind were fortunate indeed, as events would soon show.

After a weather check-out and a preflight briefing on the use of oxygen, the symptoms of hypoxia from the lack of oxygen, and other emergency procedures, the flight was ready for takeoff. Pilot Karns called Cleveland Air Route Traffic Control and notified them of the intended drop over Ortner

Airport, located on the 252-degree radial of the Cleveland VORTAC (Very High Frequency Omnidirectional Range with Distance Measuring Equipment).

Promptly at three o'clock the B-25 roared off the runway and began the climb upward with both Karns and his copilot Richard Wolfe handling the controls. Behind them, two jumpers huddled in the forward section, nine others were over the bomb bay doors and the others were in the aft section where there was a floor hatch and a waist-gunner's hatch. At 6,000 feet the pilot in charge reported to Air Traffic Control that he could see a hole in the clouds and would ascend through it, but the clouds closed around him and he requested an instrument-guided ascent up to where the blue sky was awaiting the adventurous group. This was granted.

Behind the B-25 was a Cessna 180 with a lone pilot and a skydiver. They were following the others for the purpose of photographing the jump, but the Air Traffic Center was not aware of the Cessna until almost an hour later when it reported its position over the VORTAC. At that time it was at 12,000 feet and above the cloud decks.

The pilot of this light aircraft could not at first locate the B-25, try as he might, and he circled over the VORTAC to get his bearings, intending to fly over to Ortner where the mass jump was to be made. He got one sight of the B-25 and then lost it. After getting his course and location straightened out by navigational radio, this pilot reported: "I *never saw* the B-25 after that one glimpse. I never saw the jumpers fall . . . when the B-25 said he was releasing jumpers I nosed down through the clouds at 4,000 feet per minute and broke out of clouds one mile west of Ortner Airport."

But there were no jumpers in sight. The snafu had long since begun.

Pilot Karns with the jumpers had climbed up through the cloud and haze, maintaining a heading of 262 degrees from the Cleveland VORTAC. He flew at a fairly constant speed of 145 miles an hour in order to maintain a climb of a thousand feet a minute. Upon reaching 20,000 feet, where crew and passengers were breathing oxygen to sustain life, Karns shoved his throttles ahead slightly to bring the speed up to 170 miles an hour and changed his heading slightly to the north. He was ready now to release the eager jumpers and he inched the speed downward, heading the aircraft slightly more north again, a strange move when at 20,000 feet the wind was from the southwest, which would carry the jumpers to the northeast of his position. Karns explained later that he was unable to check his navigation with the VORTAC while communicating with the Air Traffic Control Center, and he had "assumed" the winds had shifted to the north.

Pilot Karns was a busy man, there was no doubt of that . . . to speak into the microphone to the center he had to remove his oxygen mask, and there is evidence that at this time he may have been suffering from the effects of insufficient oxygen, the dangerous hypoxia. Not only that, if he switched off his contact to Air Traffic Control to check his radio navigation, he might miss instructions as well as his location, which he thought was being plotted by radar.

What he didn't know was that the Air Route Traffic Control thought the Cessna photo aircraft was the B-25 bomber and was guiding this light aircraft over the target area while, by a horrendous mistake, allowing the plane of jumpers to stray over Lake Erie.

Not only that, but there were other aircraft in the clouds below, one an unidentified strayer into the system, a TWA jet airliner, and two other airliners. Pilot Karns was ready to release the jumpers, whose hurtling bodies could rip the wings off a plane as large as the TWA's 707. The intention of the jumpers was to fall for some 17,000 feet before opening the chutes, to fall right through the clouds where the other aircraft were flying.

The conversations with the Air Traffic Controller at Cleveland, the busiest in-route traffic center on the North American continent, is preserved on tape. It is a classic memorandum of what can happen on a quiet Sunday afternoon, in skies that were reasonably busy with aircraft at that time of day. Mercifully the following events did not occur at a busier time when still more traffic was in the vicinity.

As the jump aircraft and the photographic Cessna were climbing upward to a rendezvous with death, the controller, who had watched the aircraft on his radar set and had placed a "shrimp boat" on the surface of the radar screen to identify the jump aircraft, was relieved by another controller, who saw the target aircraft was circling over Ortner Airport. Since this aircraft was not on the Air Traffic Control frequency, the controller did not continue to maintain identity, aware of the aircraft by its "shrimp boat" and over Ortner where it was supposed to be.

At 3:58 P.M. a call was received from the Cessna, advising he was southeast of Cleveland and requesting his proximity to the B-25. The controller replied: "I do have two targets out there . . . I wasn't sure whether I was watching 43-Golf [the B-25] or not, I am not sure I got him on radar . . ."

Cessna pilot: "We are heading about 260 degrees now, we can make a turn if it will help you." The controller did not accept this offer. He then proceeded to identify a westbound Aero Commander 200 whose position was then just east of the Cleveland VORTAC. This identification was required because of a departing jet from Cleveland (TWA's Flight 459 and loaded with passengers), which had previously been restricted to 7,000 feet in its climb because of the Aero Commander's assigned height of 8,000 feet. Now that the Aero Commander had identified himself, the controller freed the TWA jet to climb through the 8,000 feet and on up to 23,000 feet and the jet pilot reported okay. The controller turned to the Cessna, asked his height, and was told it was 12,000 feet.

At that moment the Cessna pilot again called and said he could see a fast moving jet below him (the TWA) and on the same heading.

Based on the position of the Cessna in respect to the big jet at that time, the controller advised that he "believed" he had the Cessna in radar contact. His testimony in regard to this use of the word "believe" was not conclusive. At one juncture he testified that he was not sure of the target's identity, but later stated that the report of the jet sighting from the Cessna "resolved what remaining doubt I had about which aircraft was which . . . because the jet had not passed the lead target yet . . . also the Cessna had to be the trailing aircraft."

The controller next called the B-25 and asked for its heading and the reply came "approximately 275 . . . ah . . . how far do you show us from Ortner now" and the controller replied at 4.01 P.M., "Looks like about three miles twelve o'clock," meaning three miles due south of Ortner Airport. Karns replied that he would be dropping his jumpers in one minute.

The Aero Commander now called with a position report and was told to omit all further position reports. Then the Cessna called and asked for its radar position in respect to Ortner and the controller replied: "I got only one target out there now . . . and I believe that's the B-25 . . . you are probably behind the B-25 about six miles. . . . The B-25 is about two miles west of Ortner now."

At 4:02, Karns called: "We're dropping jumpers right now."

The controller called the Aero Commander to say there was traffic at his "ten o'clock" position three miles away, a B-25 which had just dropped chutists.

But the Aero Commander pilot said he could see nothing ahead as "I am on instruments . . . "

Seconds later, this same pilot asked where that drop location was in regard to his aircraft and was advised that "they are off to your left rear now . . . about five miles and well clear." In other words, the Aero Commander pilot was told the men were dropping through the cloud to the rear of him when in reality the controller was watching the wrong plane and the jumpers were descending over the shore of Lake Erie. Not only that, but Karns, denied a climb higher than 20,000 feet, had never informed Air Traffic Control the men had jumped.

One minute later the controller requested the Aero Commander to report passing the Bay intersection (a radio checkpoint twenty-six miles west of the Cleveland radio on Airway Six). At that time he said: "We got a lot of clutter out there today [on the radar screen] and its kind of hard to follow primary targets [aircraft] . . . still think I got you on radar about five miles east of Bay however." Five minutes later (and no plane would be going only sixty miles an hour), the Aero Commander reported over the Bay. Even then, the Controller made no change in his identification status of this aircraft's radar target.

The next significant communication was at 4:17 and five minutes after the "Bay" transmission. Two westbound jet airliners were advised . . . "There is a Cessna out there west of the OMNI somewhere . . . we don't have him on radar . . . at around 12,000 feet . . . also a B-25 at 20,000 out in that area somewhere . . . we don't have him on radar either."

The B-25 re-established communication with the Cleveland Center as he was passing over the VOR and asked for a vector to Ortner. Seconds later he dropped the last two chutists without even notifying Air Traffic Control and with all those jets and other air traffic in the vicinity.

The first sixteen chutists fell free until they entered cloud at 6,000 feet and broke cloud at 4,500; at 3,000 feet they opened their chutes and realized with horror that they were not over Ortner Airport, but over water, Lake Erie.

They perished a few minutes later and the situation was best recalled by the last jumper of the big group. He lived because he was last out and last to open his chute and, was luckily nearer to rescue vessels on the lake. As this last jumper broke the cloud he saw a number of open canopies strung out along an east-west line descending to the water. He struck the water two miles from the others and in ten minutes was rescued. Meanwhile the two jumpers who had originally intended to jump at 30,000 feet, but had jumped at 20,000, landed where they were supposed to land, at Ortner Airport.

The tragedy disclosed so many breaches of Federal Regulations that the reader may well wonder what is going on in today's "controlled" airspace where passenger jets are flying in great numbers. Federal Regulations provide that a jump may not be made if it creates a hazard to air traffic, no person may make a jump and no pilot in command may allow a jump from his aircraft into controlled air space unless there is a two-way radio communication between the ATC and the flight;

notification must be made to the ATC of a jump one hour before the jump; all jumps are prohibited into congested areas (without special authorization from Washington); jumps are prohibited into zones under control of control towers except with authorization of the FAA; no jumps may be made into or through clouds, and no person shall make a jump less than 500 feet under a cloud, less than 1,000 feet over or less than 2,000 feet horizontally from the cloud; no pilot is allowed to permit jumps unless the above conditions exist.

Ortner Airport is located directly under Victor Airway 14, a main jet artery in the Cleveland area.

The National Transportation Safety Board concluded there was radar misidentification of the B-25 when it started on its first jump run and the radar target that the controller identified as the bomber was "most probably" the Cessna.

Not only that, the Board could not definitely account for the identity of a third target which was in the controlled area over Cleveland in addition to the TWA jet and the Aero Commander on that Sunday afternoon.

The probable cause of the accident was: failure of the pilot to terminate the mission under a condition of cloud coverage which precluded visual reference to the ground, coupled with an erroneous radar identification of the B-25 by the FAA controller which resulted in the inaccurate positioning of the aircraft. The parachutists, all of whom were experienced and aware of the hazards of jumping under the prevailing conditions, were "not without fault."

The tragedy points out clearly the fact that control over the air space in which today's heaviest traffic operates is a myth; repeated near misses and tragic midair collisions exemplify the problem which becomes more complex each day. Able now to report near misses in the air without having to give their names and thereby be fingered by fellow pilots and their union, commercial pilots are reporting double the number of serious air incidents when evasive action has been forced on them to avoid collisions.

Some of them have not been able to swerve in time. Big jets need as much as five miles to make a turn. Too sharp a turn created by an emergency can flip a jet on its back, and violent maneuvers can dislodge wing-supported engines in the twinkling of an eye. However, despite their great speed (and therefore their rate of closure upon another aircraft) and their inability to swerve suddenly, the blame for a midair collision between a TWA DC-9 and a Tann Company Beechcraft Baron on March 9, 1967, near Dayton Municipal Airport was placed on the crew of the DC-9.

In that accident, the pilot of the Beechcraft, the only occupant, and twenty-one passengers and four crew members of the passenger plane were killed and both aircraft were destroyed.

The DC-9, operating as Flight 553 out of New York City to Chicago, with stops at Harrisburg, Pennsylvania, and Dayton, was operating on Instrument Flight Rules but in Visual Flight Conditions. Descending from 20,000 feet to 3,000 feet on an FAA-controlled airway, it was preparing for the landing at Dayton. Dayton Radar Approach Control was exercising control over the flight and had made radio and radar contact with it. The visibility was five to six miles in haze.

Eighteen seconds before the collision, the radar controller advised the flight: "TWA 553, traffic at twelve thirty [almost dead ahead] one mile, southbound, slow moving."

"Roger," replied the TWA pilot four seconds later and this was the last recorded communication with the crew. The Flight Recorder Tape in the cockpit provided no information to indicate that the TWA crew ever saw the Beechcraft. Last recorded conversation on the tape was "Ready on the checklist, Cap'n." Four seconds later the recording tape stopped. The aircraft was at 4,500 feet when it collided with the Beechcraft.

The Beechcraft, on a company business flight, was en route from Detroit, Michigan, to Springfield, Ohio. The pilot was operating VFR and no flight plan was filed nor was one required. The Beechcraft was not under the control of or in radio contact with any Federal Aviation Administration Air Traffic Control facility, although the pilot was in radio contact with the fixed-base operator at the Springfield Airport just prior to the collision.

The Board pointed out that the operation of the Beechcraft was, according to the evidence, carried out in accordance with existing FAA regulations pertaining to the conduct of a VFR flight from point to point. There was no requirement for the Beechcraft pilot to contact any FAA air traffic control facility, use his transponder, which is his identifying signal to ATC radar, or display the rotating red beacon with which his aircraft was equipped.

On the other hand, the Federal Aviation Regulations restrict arriving aircraft to a maximum indicated airspeed of 250 knots when operating below 10,000 feet mean sea level within thirty nautical miles of the airport of intended landing. The flight recorder readout indicated the DC-9 was operating at a speed of 323 knots at the time of the collision, approximately twenty-five nautical miles from the point of intended landing. The excess speed contributed to the accident in that it reduced the available time for the crew of either aircraft to see and avoid the other or for the controller to take appropriate action.

In this connection the Board noted that, based on the recorder transcription, the DC-9 crew was devoting its attention to speed control, clearance response, maneuvering for the approach, and the prelanding check list, shortly before the traffic advisory was issued. This activity could direct both DC-9 pilots' attention inside the cockpit, reducing the effectiveness of any visual search for potentially conflicting traffic.

The primary responsibility of all pilots operating under VFR conditions, the Board said, is to assure that they have a clear flight path and to avoid other traffic in that airspace. The present-day "see and be seen" concept is based on all flight crews maintaining a lookout for other aircraft when they are operating under VFR flight conditions. This applies equally to the crew of the DC-9 who, although they were operating on an IFR flight plan, were in VFR conditions and were required to maintain their own lookout for other traffic in their flight path. In addition, the DC-9 crew received an accurate traffic advisory from the controller concerning the conflicting traffic, an advantage not afforded to the Beechcraft pilot.

In view of the evidence the Board concluded that although each aircraft was in a position to see and be seen by the other at a distance of approximately four miles, each of the involved air crews failed to see and avoid the other. The DC-9 was the overtaking, converging aircraft and thus was in a better position for the pilots to observe and avoid the Beechcraft. Therefore, primary responsibility for avoiding traffic within its flightpath rested with the DC-9 crew.

"The lack of positive control over aircraft operations conducted in terminal areas under the present day air traffic control system is not satisfactory," the Safety Board concluded. "It is the Safety Board's opinion that insofar as the existing ATC system is concerned, the equipment and procedures utilized by those facilities controlling air traffic may provide adequate safeguard for handling known traffic, but an equivalent level of safety is not provided when unknown traffic operations are mixed with known traffic."

One answer to this perplexing problem, the Board said, might lie in a program whereby larger segments of the navigable airspace be designated as "Positive Control Airspace," and include some terminal areas. Operations in positive control airspace normally require an aircraft be equipped and instrumented for IFR including an operational transponder and two-way radio; that the pilot be rated for instrument flight; and that the aircraft be flown under IFR at a specific flight level assigned by ATC.

The Board also pointed out that a practical Collision Avoidance System (CAS), now under experimental development, suitable for use on the majority of aircraft, would provide a great contribution to flight safety. Such a system would detect a potential collision hazard, call the pilot's attention to the hazard, and display the evasive action required by the pilot to avoid a collision.

Subsequent to the accident, the Board recalled that the FAA, in an action aimed at helping prevent similar accidents, issued, in August 1967, Advisory Circular No. 90–32, which informed the aviation community of the capabilities and limitations of radar systems and the effect of these factors on the service provided by Air Traffic Control facilities.

As a possible additional safety measure, the Board said, the FAA is also studying the feasibility of climb and descent corridors for use by high performance aircraft at major air terminals.

The reader, as well as every air traveler in the country, might well wonder why a general aviation aircraft four miles in front of a descending jet with passengers should have been permitted in the air corridor in the first place, "just in case." The problem will soon have to be solved because the public, by and large, does not believe that general aviation and commercial aviation should use the same airspace or the same airports. Unless these two sorts of aviation are separated, even more midair collisions will result.

Just why the crew of a regularly scheduled flight into a large city municipal airport, and on its final descent with a crew preoccupied with the complicated descent check list, should have to try and locate a small aircraft in haze and avoid it is beyond understanding. Perhaps litigation in this case may find that the courts think differently from the Safety Board.

On Sunday, August 4, 1968, a North Central Airlines Convair with eleven persons aboard collided with a small engine aircraft while approaching Mitchell Field, Milwaukee, in a controlled airspace under the radar guidance of the FAA Approach Controller. The force of the crash buried the small plane, with its three dead passengers, in the passenger aircraft. The pilot of the North Central skillfully flew his aircraft to a safe landing. How many more incidents like this will be tolerated? Perhaps many, because the general aviation lobby is a powerful force already complaining of commercial airliners who hog the airspace and force them to wait in long line-ups for takeoffs, endless circles before landing, and watchfulness that was never needed prior to the jet decade.

Someday, all commercial aircraft will be separated from noncommercial both in the corridors in which they fly and the airports that they use. The increased use of high-speed jets for business travel has further complicated the air traffic controlled space, and in the future these aircraft also may have to be linked with general aviation in the assignment of airspace with a computerized network to guide them.

Yet, strange as it seems, the great midair collisions have occurred in controlled airspace and, almost without exception, under radar surveillance. The classic example was the collision between two airliners over the Grand Canyon away back on June 30, 1956, when 128 persons lost their lives. More recently, on December 16, 1960, a United DC-8 jet and a TWA Constellation collided over Brooklyn resulting in the death of 116 passengers and 12 crew members. Thirty-six persons died in Canada when a TCA airliner collided with a small air force plane in an assigned airspace in Western Canada. On February 8, 1965, an Eastern Airlines flight crashed into the Atlantic Ocean while approaching John F. Kennedy Airport and 84 persons died and the probable cause of this accident was the evasive action taken by the flight to avoid what the pilot mistakenly thought would be a collision with a Pan Am 707 jet inbound from Puerto Rico.

Soon after this, on December 4, 1966, a TWA 707 jet collided with an Eastern Constellation over Carmel, New York, at an altitude of 11,000 feet while approaching New York's Kennedy Airport. Four persons died, and the pilots of both aircraft saved their passengers only by flying their damaged planes to skillful landings, one a routine airport landing and the other a crash landing below where the collision occurred.

Miraculously, 108 persons in the two planes escaped death, but the captain of the Constellation, forty-two-year-old Charles J. White, sacrificed his life after bringing his aircraft to the ground by rushing into the fire-filled aft section to assist a passenger who was in difficulty. They perished together. The cause of the collision was said to be a misjudgment of altitude separation by the Eastern crew because of an optical illusion created by an upslope effect of cloud tops. This led them to take an evasive maneuver which resulted in the collision.

These two aircraft were being funneled into the world's busiest air corridor with only one thousand feet to separate them. There is always the possibility of altimeter failures or misreadings (which are common) and, in this case, thick cloud to compound the situation. The situation is one that is serious at the moment and getting worse. Midair collisions seem to occur only in the United States traffic system and rarely, if ever, happen outside this country. Perhaps the rate of incidence is related to volume, and surely computers could soon establish the highest incidence period of midair collisions. Evasive tactics could then be implemented, if there is any airspace left into which a turn could be made.

On March 28, 1967, two aircraft collided over Long Beach, California, in clear weather. The Safety Board found that both pilots failed to see the other and perhaps sun glare was the contributing factor. Luck seems to play an important part in today's flying. Some traffic controllers would like to have solid cloud at all times and have everything under positive control. They may get their wish someday.

It is interesting to note that after the Cleveland parachute mishap—and perhaps it is just coincidence—the cross-country jet flights and especially those from Chicago and Detroit to New York City, were being taken over by the Toronto, Canada, Center. It was already handling heavy transatlantic traffic from Los Angeles, Chicago, and Detroit as well as its own international and domestic operations. But the skies were clearer of traffic over southern Ontario and this switching of airspace control was all to the good.

But Canada has its own problems, particularly in the busy Montreal area. There is one spot, west of Montreal's International Airport, which pilots refer to as "death valley." It is a checkpoint over Hudson Heights that aircraft inbound and outbound from Montreal must often cross—a common radio reporting center that puts too many aircraft too close together when it is not necessary. But still it remains, despite the protests of the Canadian Air Line Pilots Association. The protests are well founded. An aircraft flying from Toronto to Montreal will keep to a southern route, crossing over Massena, New York, and then sweeping downward to cross over Hudson at 6,000 feet, cross the Ottawa River and fly north of Montreal Airport. Ten miles east of the airport, it will swing south and west and come in for landing, keeping Mount Royal to the south and yet flying over busy Cartierville Airport at a thousand feet or less.

But the problem is back at Hudson. Jets and turboprops taking off into a west wind from Montreal fly to Hudson under 6,000 feet before getting permission to climb up, cross over Ottawa and then head to Toronto. They turn south at a radio intersection called Klienburg, then proceed right over the heart of the downtown skyscrapers to approach Toronto International Airport over high-rise apartments and solid subdivisions.

The hairy part of the trip is the crossover at Hudson, where jets are separated . . . westbound under 6,000, eastbound at 6,000, both often in the thick cloud so familiar to the confluence of the Ottawa and St. Lawrence rivers at this point. It is frightening, to say the least. The simple solution would have been to route all westbound flights over Massena and all eastbound over Ottawa and there would be no crossover at the one radio point, but this seems too much to ask. Canadian and U.S. pilots are waiting for the awful day when one of them will smack into the other over Hudson. It seems unfortunate that guide rules must be determined by accidents and by multimillion-dollar litigations.

One such battle is now in progress as the result of a midair collision on July 19, 1967, near the airport at Asheville, North Carolina, when a Piedmont 727 jet collided with a small plane shortly after taking off. Eighty-two persons lost their lives. In answer to the first of many million-dollar lawsuits over the crash, Piedmont Airlines asked Federal Courts for dismissal, claiming that all suits be directed against the U.S. Government because of "negligence of Federal employees in the Asheville airport tower" which was "one of the causes or the sole cause" of the disaster.

Shaken by the increasing number of midair collisions and near collisions, the Federal Aviation Administration found it necessary in April 1968 to release a survey on the situation, part of a year-long 1968 survey which will be made public sometime in mid-1969. The preliminary report was enough to send chills up the spines of seasoned air travelers.

In the first two and one-half months of 1968, FAA received 554 reports of near midair collisions from pilots and other sources. FAA estimates that during the reporting period there were approximately fifteen and a half million flight operations of all types made in the United States. At this rate, one near midair collision report was filed for almost every 28,000 operations.

Of the 554 near midair collision reports filed as of March 18, general aviation pilots submitted 251, airline pilots 160, military pilots 141, and air traffic controllers two. Incidents in the terminal area were the subject of 339 reports. The remaining 215 dealt with en route incidents.

"In order to encourage frank and complete reporting of near midair collision incidents during the course of the study, FAA will not take any enforcement or other adverse action against any person filing a report, even if investigation discloses he has violated pertinent operating regulations. As an additional inducement, the agency will not reveal the identity of anyone reporting an incident or the details surrounding the incident," it was announced.

To date, the study group has isolated ten broad areas for further study: mix of aircraft operating under instrument (IFR) and visual (VFR) flight rules, airspace navigation problems, the see-and-avoid concept of maintaining aircraft separation, training flights, pilot deviation, mix of high and low speed aircraft, airport traffic patterns, air traffic systems errors, proximity of airports, and marginal VFR weather conditions.

Reports also are being analyzed to determine the degree of hazard involved in each incident. Of the 436 reports (those submitted in January and February) on which analysis has been completed, 250 were classified as representing "no hazard" to either aircraft, meaning that the path of flight was such as to assure separation between the two planes without the need for evasive action on the part of either pilot. The rest—186—have been classified as "hazardous" situations; 117 occurred in the terminal (airport) area and 69 in the en route area.

In contrast to the terminal situation, 33 of the total of 69 en route incidents, defined as "hazardous," involved aircraft in which both planes were operating VFR (eyes open) and 25 incidents involved both instrument and VFR aircraft.

Complete automation seems to be the only answer to the complex problem of safety in the air. The automated air traffic control system is capable of displaying, on FAA controllers' radar screens, luminous data blocks of flight information. The data blocks, called "alpha-numeric tags," contain coded flight information in the form of letters (alpha) and numerals (numerics). They also may contain symbols. Alphanumeric tags are electronically "attached" to the correct radar blips and thereafter automatically follow them across the radar screens.

Flight information displayed in alpha-numeric tags may include aircraft identity, altitude, attitude (whether climbing, descending, or in level flight), and other data needed in air traffic control. Automation is designed to take over much of the routine bookkeeping work now performed manually by air traffic controllers.

In today's system, by comparison, the controller must either memorize the identity of radar blips or write coded information on clear plastic markers. He then moves the markers manually across horizontal radar screens. Keeping the marker near the moving blip helps the controller remember the correct identity of individual radar blips, which often number in the dozens during moderate to heavy traffic periods.

Also, much of the supplementary flight information, such as altitude and time estimates along the route of flight, currently must be handwritten by the radar controller on paper "flight progress strips." The strips are placed in indexes alongside radarscopes. Thus, the controller's attention is diverted from the scope each time he refers to or revises his bank of flight progress strips.

The semiautomated system, on the other hand, puts most of the vital flight data in the alpha-numeric tag—on the same radarscope that shows the second-to-second progress of flights. It is anticipated that automation eventually will eliminate the need for flight progress strips altogether.

For aircraft equipped with automatic altitude reporting transponders, the alpha-numeric tag includes a continuous numerical readout, similar to an automobile's odometer, showing the actual altitude of the aircraft at each 100 foot level. In today's system, controllers view only a two-dimensional (range and bearing) radar picture of each plane's position. The third dimension—altitude—must be radioed by voice from the pilot to the ground-based controller.

The semiautomatic system also performs many other vital functions in the air traffic control system. They include: automatic coordination and transfer of flight control between controllers and between adjacent control facilities; automatic updating of flight information; automatic printing and distribution of flight data to control sectors; error checking of pilot's and controller's actions; automatic processing of flight plans; and electronic display of significant in-flight weather conditions.

Eventually, automation by 1970 at major airports will take on such additional air traffic control functions as predicting impending traffic conflictions and suggesting ways to resolve them, flow control advice in congested traffic situations, and preplanning the sequence of airport arrivals.

In London, England, "hands-off" landings have been in practice since early 1968 with excellent success, but this system may not be introduced for some time into the U.S. airport complex. When it does come, landings and takeoffs will be speeded up and flying may become safer. But its success may depend on the pilot of a small plane without proper navigational equipment wandering through cloud near a big city airport, listening to a ball game and looking for a railroad track.

Control will have to mean CONTROL over every flying machine in the air in every segment of the country.

12

Airport Is a Dirty Word

Time was when an airport was a respected member of the community. Today, nobody likes an airport. It depreciates land values, curtails suburban expansion, pollutes the air, irritates both pilots and passengers, creates traffic problems on the ground as well as in the air, and is a constant menace to surrounding communities because almost all air disasters occur in its immediate vicinity.

The airports that were so familiar in the early days of flying have gradually disappeared and the level grass fields have been swallowed by the expansion of suburban growth. Airports that once were in the country, protected by farms, lakes and trees, have been encroached upon by an endless deluge of new housing projects, high-rise apartments, and small industries. Attempts to relocate airports farther in the country are met by a barrier of militant farmers and rural colonists. Lengthening of airport runways for safety and the expansion of terminal and hangar facilities meet with riotous reaction by otherwise peaceful and law-abiding citizens who are not concerned one whit with air safety but are fed up, and rightfully so, with kerosene fumes and the thunderous noise that keeps them awake, disturbs their schools, reduces their religious services to unrecognizable mumbling, and leaves them apprehensive about the security of their home investment and their lives.

It would be simple to blame today's airport mess on bad planning, and bad planning should certainly take its share. But the problem lies much deeper than that. Commercial air traffic and general aviation have mushroomed so tremendously that landing space for both is disappearing fast, and suggestions for relocating commercial airports far out in the country are met not only by the militant rural inhabitants but also by hostile passengers who argue that already more time is spent in ground transportation than in flying from one city to another. Pilots complain that landing and takeoff flying procedures to reduce noise and pollution over crowded housing developments are often dangerous and the Federal Aviation Administration, in agreement, has ordered all pilots to fly their planes under acceptable procedures only, and to ignore municipal interference. This decision has created thousands of lawsuits and millions of resentful citizens. One airport had over three thousand lawsuits pending in 1968.

What is to be done?

The solution of this monumental problem appears to require an entirely new approach by the public, the flying traveler, commercial and general aviation pilots and owners, the airlines, local governments, and Federal agenices who, together or separately, will have to shell out some five billions in the next five years to bring order out of the chaotic conditions that exist today. Airports are suffering from too many planes, too many people, and too many automobiles. There are tremendous traffic jams on the access roadways and in the parking lots in terminal space. Baggage systems, restaurants, washrooms, lounges and access corridors all suffer from the overcrowding that is threatening to engulf the system as it now exists.

It is a cruel and unfortunate paradox that the country is planning and constructing bigger and better office buildings, larger and more luxurious homes, more freeways, larger schools, modern libraries, university cities, and planned metropolitan satellites of the future, yet is losing the battle of airport progress. What is left at the moment are the familiar congested centers that were constructed solely for the convenience of the airlines, leaving harassed passengers to walk two thousand feet or more at every major airport in the country, while at

Chicago's labyrinthine O'Hare Airport they must slog nearly a mile to get to an airplane. The Federal Government, which is showing leadership in the construction and financing of new airports, is at a loss when it comes to helping surrounding communities to build access highways to handle the airport ground traffic. The government looks upon airport access roads as only one ingredient in any overall metropolitan transit systems. As a result, communities shelve all proposals of building airport roads as unnecessary waste of public funds for the 40 percent of the population who flit about the country.

Helicopter transportation, often thought to be a solution to the access problem, had a severe setback in 1968 with heavy loss of passenger life in Los Angeles; this put the crimp in any expansion in this direction. Some airport authorities believe the answer is to go up, up, up, and build tremendous skyscraper parking lots with the terminals directly below them to save further passenger inconveniences. But this was tried in Toronto, Canada, and instead of the new terminal servicing the air passengers, it became a center for visitors and spectators and the air travelers were forced, and are still being forced, to park in open fields and trudge a mile or more to the passenger counters.

Chicago's authorities have suggested building a new jet airport out in Lake Michigan. The cost would be horrendous and the location would not solve the access highway problem, though it would help defeat noise and pollution. In case of a bad landing or takeoff, passengers would be equipped with lifejackets and inflatable rafts.

Los Angeles would like to go underground, using the vast space beneath the runways for terminals, parking and maintenance buildings—a good idea but one that does not solve the safety, noise, and pollution problems. Pollution has now become a major nuisance when you consider that every time a four-engine jet takes off it dumps more than seventy-five pounds of solid pollutant into the atmosphere. In 1968, meteorologists found that polluted air blends quickly into the manufacture of raindrops to create more violent storms as well as surface wind conditions with the velocities of hurricanes. Yet these atmospheric disturbances are not predictable by present methods of forecasting.

Suppose Los Angeles does go underground. How can nearby communities be prevented from permitting the construction of high-rise apartments that impinge upon the approach to the airport. Not long ago, the city of *Inglewood,* for instance, enacted an ordinance permitting high-rise buildings in certain areas, one of which intrudes on the approach pattern of Los Angeles Airport. The Federal Aviation Administration may try to curtail such a move by declaring it a hazard, thereby complicating the construction loans and insurance. Nevertheless the FAA would be unable to prevent the move by direct intervention. Such is the weakness of a national transportation system that is hamstrung by lack of enforcement machinery.

The problem at Los Angeles is basically the same as it is everywhere else, except that Southern California has the busiest airspace in the country when general aircraft is added to commercial operations. Though Los Angeles is the fifth largest airport in the country and nearby Van Nuys is the busiest general aircraft airport in the world, the operation of all the airports in the state remains under local community control instead of under a broad airport authority. As a result of local control, the Los Angeles area has allowed sixty-three airports to dwindle to nine over the past few years and has been unwilling to give zoning for a single new airport in more than two decades.

There are some who argue that freeways were imposed on the people, why not airports, forgetting that airports and airlines can be sued to the hilt for causing loss of sleep, daytime nervousness, dirty washing, crop damage, and also destruction of mink breeding, among other things.

One of the world leaders in airport administration and planning is Francis Fox, general manager of the Los Angeles Department of Airports. With architects and engineers nodding approval, Fox says that the major items on his planning front include the development of the present International Airport to its ultimate capacity, creation of a master airport system for all of Southern California, and a solution to the "ground barrier," which is what he calls the travel time to and from the airport by air passengers.

His idea embraces the entire spectrum of airport development and should be a guide to all other major airport cities in the country. It will require an improvement in the so-called clear zones around the airport, expanding the existing decentralized terminals, planning new terminals, enlarging of airport runways and roadways, increasing auto parking by multilevel parking structures, expediting pedestrian flow from parking areas to terminals, developing a cargo city, and increasing the gate positions from the present sixty-eight to more than one hundred.

When the project is completed, Los Angeles will be capable of handling up to 30,000,000 passengers annually, instead of the current 13,000,000 with their 46,000 cars in and out each day.

By 1975, Fox anticipates there will be 90,000 cars daily, and he sees the saturation limits of the airport taking place in the late 70s or early 80s, when cars will number over 100,000 daily and passengers will increase to well over 35,000,000 a year. The capacity of Los Angeles could reach saturation between 1975 and 1980, but there are those who believe it will peak before then because of the projected 500-passenger jumbojets which could provide a round trip to Hawaii for fifty dollars. Then Los Angeles will be inundated despite the planned improvements.

The situation at New York is abominable. Kennedy International Airport on Long Island (once believed to be the "unlimited airport"), Newark Airport in New Jersey, and La Guardia in the very heart of the borough of Queens are congested and dangerous; yet attempts by the Port of New York Authority to purchase land for a giant superairport to meet the needs of the jet age have been shot down by every state and community within driving range of the great metropolis.

The density of aircraft over the New York complex would have long ago reached saturation, if the FAA rules on aircraft separation had been scrupulously followed by Air Traffic Control. To show how serious it had become, controllers in July 1968, angered at Federal budget cuts which would have deprived them of pay raises, returned to "regulation" flying procedures. This "slowdown" was actually created by the controllers "following the book" on three mile aircraft separation in the New York approach zone and it shook the government into action. Some flights were delayed as long as six hours. This was particularly irritating to International flights which had to circle endlessly, it seemed, after long trips from overseas. Flights from Montreal, Toronto, Detroit, Cleveland, Washington, and elsewhere were held on the ground in those cities for as long as three hours while the controllers "brought 'em in by the book."

What may have been overlooked in the snarl and the recriminations that followed was the fact that the FAA was permitting controllers to abandon the safe flying rules because saturation would already have been reached if they were enforced.

But if the air travelers thought this was a single upsetting incident caused by a wage dispute, a look at New York's future is interesting. The "hub" is now introduced to refer to a complex of airports that feed one community area. In the New York hub in 1965, for instance, some 12,325,000 passengers were served. By 1970, this figure will jump to 22,464,000; by 1975 it will escalate to more than 38,000,000; and by 1980, just over a decade ahead, New York hub airports (wherever they may be located at that time) will serve or hope to serve 64,469,000 passengers—some 16,000,000 more than Chicago will serve at that time.

The complexity of New York air operations necessary to fly so many passengers is brain staggering. There were 2,389,000 aircraft movements in 1965; by 1975 this will jump to 5,500,000; and by 1980 up to over 8,000,000. The flights of general aviation will double the number of commercial operations at that time.

It is all very well to say "chuck the small private planes and business aircraft operations," but these flights today make up a sizable business force with corporation jets whizzing through the air on business flights, small single- and twin-engine private aircraft also on business and some on pleasure. The total makes quite an economic impact. Not only that, but many corporations and small business firms, seeking to expand beyond the big cities invariably look for communities that are served by airports that encourage general aviation. The battle lines between commercial and general and military have been drawn and fought over, but no one knows how to end the battle.

The Federal Aviation Administration is all too aware of the complexities of the airport problems. In the Second Jet Decade, over seven thousand turbine-powered airplanes will be flying and they will have to blend somehow with the increasing but slower fleets of small aircraft. The bigger aircraft will naturally fall into reasonably fixed geographical areas. In 1965 (the base year we have been using), twenty-one metropolitan areas in the United States served two-thirds of air line passenger enplanements. In the next decade they will handle over 70 percent, and maybe more, and every plane in the air will be flying faster.

Revenue passenger miles will increase from 76 billion miles of the base year to 315 billion passenger miles by 1980, while cargo will jump from approximately three billion ton miles to over thirty-eight billion ton miles, and the aircraft fleet to handle it will multiply appreciably. There are those who think that cargo aircraft will far outnumber commercial jets by the end of the next decade. It has been estimated that the 97,000,000 flown in the year 1965 will be increased, as early as 1975, by an additional 240,000,000 passengers.

As an indication of just how big is big, United Air Lines alone carried 18,040,000 passengers in 1967, equivalent to the combined populations of New York, Chicago, Los Angeles, Philadelphia, and Baltimore, and the company's marketing experts expect this traffic to double every five and a half years. To meet this demand, United and other U.S. flaglines will spend some eighteen billion dollars in a ten-year period for aircraft and modernization and enlargement of ground facilities to meet the crush. Can safety standards possibly keep abreast of this tremendous increase when they have been unable to do so in the first decade of the jets?

William F. McKee, former administrator of the FAA, claims there are three essentials necessary for the safety and growth of aviation: (1) an adequate system of airports; (2) an automated air traffic control system; (3) trained personnel to operate and maintain the system.

"The inadequacy of our airports is the greatest single limitation on the capacity of the nation's air transportation system," he said. "Many of our major airports are suffering from insufficient capacity. Aircraft cannot land or take off at a rate which the traffic demands. Many of the constraints stem from sheer lack of concrete, in the form of runways, taxiways, ramps, and so on. The number and duration of these delays is serious. In a 1966 FAA study of delays at 304 terminal areas covered by control towers, [delays] amounted to 173,000 hours. These delays cost the domestic airlines $57,000,000."

McKee claims that over 2,000 of the nation's 3,200 publicly owned airports now in use must be improved to handle the rush of the next decade. Needed to augment these existing airports will be 900 additional airports of which 225 would be located within the hub areas of the big metropolitan complexes. They will be "reliever" airports, but added to these will be an additional thirty-five supermodern airports able to handle the supersonics, the jumbojets, and the demands of all the commercial carriers as well as filling the needs of the citizens who never fly at all.

The FAA recognizes the mess. So does Alan Boyd, the very capable recent Secretary of Transportation, who is trying to force greater safety into all aspects of air transportation because of the knowledge that the future indicates a jam-packed sky, where the human element collapses and only computers can solve the muddle.

But although computers may solve the landing and takeoff frequency in the next decade of jets, high costs of equipment will prevent the continuance of private flying. This means (although the government agencies don't mention it) that the time is soon coming when all pleasure and business flying will be separated from commercial operations entirely.

Meanwhile, the industry is well on the way to providing a dependable collision warning device system for use on large aircraft only, more towers, more radar, and more instrument systems. Yet control towers are lacking in 235 airports which now serve commercial purposes. Such towers must be erected soon.

Also needed: the expansion of air route traffic radar service with an estimated 140 new long-range sets to be installed by 1978; 850 new instrument landing systems at major airports; special landing facilities for helicopters and vertical takeoff types of aircraft; and other proposals including taxes to cover the costs.

The Air Line Pilots Association, appearing before a government inquiry in August 1968, stated in a brief that "some period of time will be required until improvements are made in Air Traffic Control and airports. . . . The air traveling public and cargo shippers will suffer delays and the air carriers economic losses due to these delays . . . it must be recognized that the present delays are creating a hazardous situation due to the pilot being unable to get reasonably accurate expected approach clearance times and are often being misled."

Pilots are of the opinion that airport delays are not so much due to regulations and procedures as to the fact that the airspace in high density areas is not used efficiently enough to provide the requisite flow, nor are there enough landing aids and runways to satisfy the peak hour demands.

"Substantial immediate relief can be obtained by scheduling airline aircraft so as not to exceed the instrument capability of a particular airport and its environs and attracting General Aviation to satellite airports," the association reported.

The pilots were also of the opinion that some relief can be obtained immediately by decreasing radio frequency congestion and consequent controller workload by the elimination of radar vectoring except for separation purposes, declaring a moratorium on all controller duties not directly related to traffic control, requiring aircraft flying into high density airports to have minimum equipment such as two VORs (radio range navigation equipment), two crystal controlled transceivers, DME (Distance Measuring Equipment), ILS (Instrument Landing System) capability and transponders (positive aircraft location established on the airport radar screens).

The ALPA would also require all pilots flying into high density airports to have instrument ratings and take part in all current briefings regarding the latest in procedural requirements. If these guidelines were enforced, most general aviation would be eliminated from the big airports. The "casual Sunday fliers" would be confined to other airports where their possible poor flying grades, inability to navigate without a railroad track, inability to have or read radar in machines ill-equipped for high density operations, and where the occasional gulp of alcohol would not endanger the lives of hundreds of travelers.

ALPA believes that augmenting computer equipment to give each pilot better area navigation capability would relieve the Air Traffic Control congestion and this would be possible "without the expenditure of any further government funds."

The pilots claim that all these proposals could be in use by mid-1969 and the efficiency of the Air Traffic Control system in the airport vicinity as well as along the high altitude routes could be increased by the introduction of the data linking of communications, more refined area navigation for the thriftier use of airspace, airborne collision avoidance system as a cross-check on the accuracy of the system, and the use of a speed control system over all aircraft to adhere to the flight plan—in other words a control that would keep each jet at a particular takeoff, travel, and landing speed so that the controllers would be able to follow and advise every flight by mathematical computation. It would, of course, limit the speeding up of flight times to meet scheduled arrivals due to late departures and the everpresent challenge of making records between cities.

In one crash, investigators found a printed slogan in the wrecked cockpit: "Beat American."

If the pilots have their way, the skies of the future will be filled with controlled aircraft flying at prearranged speeds, taking off and landing under precise speed controls. But whether or not the airlines and the public will buy all this remains to be seen.

No sooner had the ALPA issued its recommendations than the House Appropriations Committee made deep cuts in the FAA annual budget for 1969, eliminating the customary $75,000,000 grant for airport development while maintaining the necessary $663,000,000 for operations after the hullabaloo created by the New York Center slowdown—timed to the moment to teach the Appropriations Committee a lesson.

It is too bad that the entire deteriorating airport situation did not get the same publicity. It is a disgraceful mess which only Federal intervention can remedy, and that is unlikely no matter what the toll in time and life.

At the present time there are approximately 567 airports in the United States with daily commercial scheduled services. Of this number, some 217 are totally inadequate in rescue and fire fighting services; 141 others have been rated as below standard by the Air Line Pilots Association; while another 95 have only token equipment available.

It is beyond belief in this age of modern transportation that, if a two-minute time response for fire fighting equipment at the country's major airports had been used in the survey, no more than twenty-five of these airports would have remained in the "A" category.

In other words, as far as keeping abreast of the present needs of safety at the country's commercial airports goes, only twenty-five airports fill the bill. The rest fall into categories that reveal a widespread lack of facilities to cope with many emergencies.

ALPA surveyed each airport generously, too generously in the opinion of some. The organization gave an "A" rating to airports with adequate equipment and manpower; a "B" rating to those below standard in equipment and manpower or both; "C" for only token equipment; and "D" for no equipment whatsoever. To know that 217 commercial airports are in category "D" is enough to give any air traveler the shakes.

After all the crashes at the Greater Cincinnati Airport, one would think that the airport would have an "A" rating. Wrong. It is only "B"—below standard in rescue and fire fighting equipment as of mid-1968.

Nearly all large and hub airports having four-engine jet service have fallen into either "A" or "B" categories, but two airports with such service fell into "D" class. They were Shreveport, Louisiana, and Palm Springs, California. Shreveport had forty-two scheduled flights daily in 1968, while Palm Springs had fifteen commercial flights and a veritable host of general aircraft flights. Neither has fire equipment of any kind . . . not even a hand fire extinguisher.

The following airports with little or no fire equipment or rescue services, and yet using three-engine jets, are: Sitka, Alaska (D), Santa Barbara, California (C), Fort Myers, Florida (D), Pensacola, Florida (D), Sarasota, Florida (C), Rochester, Minnesota (D), Bismarck, North Dakota (C), Grand Forks, North Dakota (C), Columbia, South Carolina (C), Chattanooga, Tennessee (C), Corpus Christi, Texas (C), Lynchburg, Virginia (C), and Roanoke, Virginia (D).

In the State of Virginia there are ten commercial airports with frequent daily operations, but Roanoke seems to be the classic example of inadequacy. This airport has some fifty average daily services from Eastern and Piedmont Airlines and yet has no fire fighting equipment or first aid and rescue services whatsoever. In the entire state there is only one "A" category airport—Norfolk, which has only three more daily flights than Roanoke.

A quick look around the country shows the current state of the airport situation. Mobile, Alabama, has twenty-two flights daily and is rated "C," in Arkansas, the cities of Hot Springs, Jonesboro, Pine Bluff, Texarkana, and others are rated "D," San Jose, California, is only a "B," while the modern city of Santa Barbara can manage only a "C." Of the eighteen airports in Florida where some of the nation's heaviest commercial and general aviation flights are recorded, only eight are fully adequate, three are below standard, three have only token equipment, and four have no fire services.

The great airport of Wichita, Kansas, where four major commercial air lines operate gets only a "B" category by the pilots, while Lexington, Kentucky, gets a disgraceful "D."

In all Michigan there are only four airports with a top rating: Alpena, Metropolitan Detroit, Detroit Willow Run (no longer used by commercial air lines), and Traverse City. Both Washington, D.C., airports are top-rated and three out of four in New Jersey hit the ultimate, showing it can be done, But in busy New York State, only ten airports get the "A" while eleven others are ranked lower. In North Dakota there are no "A" categories. Of the twelve major airports in Wisconsin, two have top rating, there are no "B" categories, and all the rest are in "C" or "D," with seven having no fire equipment at all.

But if the Air Line Pilots are justly concerned about the lack of fire fighting equipment and rescue services at the nation's commercial airports, they are equally concerned over the lack of radar surveillance systems, instrument landing approach systems, and the general inadequacies of airport runways both in length to meet the requirements of the jet age and the lack of overruns and underruns—hard surface extensions of the runways to provide safety during too early a descent or an aborted takeoff.

A mid-1968 survey of the airport situation revealed an unbelievable lack of control towers, radar, instrument landing systems, lighting, and round-the-clock airports with "power" turned on. For instance, the State of Idaho, with eight commercial airports, has radar at none of them, control towers at only three, and approach lighting at two.

Florida has seventeen busy airports with particularly heavy traffic during the winter months. Yet nine of the eleven jet airports have runways below the FAA recommended length for safety. Only Fort Lauderdale has a runway longer than required and Miami International Airport has a 10,500 foot runway which is regarded as merely adequate.

Busy Atlanta, one of the most congested airports in the country, operates with the main jet runway 1,800 feet shorter than the FAA recommended length, while Chicago's major airports are longer than the existing requirements but have insufficient overrun and underrun areas.

In the entire state of Kansas there is only one airport with radar surveillance, although there are twelve commercial centers. In the modern state of Michigan there are twenty-four airports but twenty-two have no radar and thirteen have no towers. On top of that, the two major airports of Detroit Metropolitan and Flint have jet runways below the FAA requirements for safety.

At greater Cincinnati Airport, where so many approach crashes have occurred over the jet years, the jet runway is only 8,600 feet long, some 2,000 feet short of the recommended government length, and at busy Cleveland the main jet runway is 1,200 feet under the recommended length. On the good side, Columbus has a 10,700 foot runway, more than 2,000 feet longer than is required. Runway lengths are decided upon by FAA and ALPA experts, taking into consideration the type of operation and the approach and takeoff hindrances. Anything less than recommended runway lengths is courting disaster, and there is no doubt that runway lengths will be the basis for litigation in future accidents on airports.

New York State boasts twenty-two airports in commercial service, of which twelve are jet airports. Yet twelve lack radar and nine lack tower control, while Buffalo, Rochester, Utica, White Plains, Massena, and Newark (servicing New York)—and most others—have runways that are below Federal standards. North and South Dakota have no radar at any of their airports. Neither has mountainous West Virginia, while Roanoke, Virginia, has a jet runway that is only 5,427 feet long when it should be not less than 6,400 feet. But then big complexes like Philadelphia are a thousand feet short and busy Pittsburgh is 1,300 feet short by FAA standards.

It's bad enough having insufficient runway lengths at commercial airports as well as lack of radar, control towers, and instrument landing systems. When one thinks that of the 567 commercial airports from the Atlantic to the Pacific, 334 of them have no approach lighting, one begins to wonder where all the billions were spent that went into the National Aviation and Space Administration through the jet decade. This organization is supposed to take care of the needs of both aviation and space. It has permitted aviation progress to lapse in order to launch the space effort, because of recurring propaganda that "the Russians are ahead of us." No wonder the American public is bewildered.

Not all airports are publicly owned. There are more than ten thousand airports in the country and two out of three are privately owned. With the exception of the large commercial centers, general aviation pays a substantial part of the operation of these airports, paying some fifteen million dollars in 1966 for Federal fuel tax alone. It is the opinion of general aviation that air communication between small cities, towns and villages and between private industrial sites must be accomplished by them, which means that although all jet commercial flights are above 18,000 feet and most general aviation flights are under this altitude, there must be overall control of all these operations at all heights, in all districts of the country, to bring airports in line with Federal standards and flights under computerized control.

Any suggestion of regulation over general aviation always brings a great howl from the Owners and Pilots Association, but the airport problem is so complex and so necessary to social and economic growth that general aviation must receive a just share of public monies for small airports beyond the great hubs. By 1975 there are expected to be some 250,000 general aircraft flitting about and facilities must keep pace. There are 3,830 publicly owned airports at the beginning of 1969, and approximately 3,240 privately owned airports not open for public use but nevertheless generating business and training fledglings.

All these airports are used by general aviation and this flow of aircraft also uses commercial airports for feeder flights, business and pleasure flights, and in some cases for training flights. If general aviation is a menace to big commercial operations at present airports, then modern radar-controlled and well-lighted general airports should be constructed or old ones rejuvenated, to ease congestion at big city airports, speed up commercial operations, maintain a higher degree of safety by separating jets from pistons, and give general aviation a proper place of operation instead of the helter-skelter system or lack of system now in use.

This needs programming at a much higher level than local control. Take the case of Pontiac, Michigan, a modern city in the Midwest once served by commercial planes but too close to Detroit and to Flint to support such operations. Pontiac is bulging at the seams with general aviation, and airport enlargements at the Pontiac Municipal Airport have been gnawing into suburban housing and recreational areas.

Surrounding Pontiac and not more than five miles in any direction are four other airports, three of which are jammed with small airplanes. One is a privately owned strip with jets operating from it. Directly above all these airports jet airliners descend toward Flint, Windsor, and Detroit Metropolitan airports; the skies are crowded with every sort of aircraft imaginable, including big jets and transports from Selfridge Air Force Base. The air is congested, not because of a lack of planning, but because these airports were once in the back country and the cities have overrun them.

Faced with the problem, Pontiac planners decided on a jet airport for the future. Land was purchased for the project. Then someone pointed out that the new airport was within a mile or so of a mushrooming complex of planned homes and retail businesses that may soon explode into the first totally planned city in the country. The matter stands at the present because of local squabbles, but Pontiac needs most what the entire country needs—a Federal Airport Authority which would advise, plan, and assist in the building of new airports, consolidating older ones, and planning the access road systems and rapid transits to serve the airports.

Not only small communities become bewildered. Large ones too feel the hot breath of confusion. The Port of New York Authority which operates many forms of transportation facilities besides airports, has been criticized as having done less to solve its present and future local transportation problems than any other city in the world.The blame has been attributed to political bickering.

In Miami, the Dade County Port Authority, which can cross county lines, has been attempting for years to unjam the airport approach roads by trying to build access routes of its own but political moves have blocked the project in a sort of running battle, not unlike an old-time movie serial. Rapid transit systems that might solve the problem have, for the most part, bogged down in long and complicated studies overshadowed by costs. A shining exception is the Cleveland Transit System which will connect the downtown to Hopkins International Airport with a twenty-two minute hop on a Pullman stainless steel train of twenty cars, each with a capacity of eighty passengers. In London, five-hundred-seat passenger trains will soon whisk air passengers from Victoria Station to all three London Airports, Heathrow, Gatwick, and Stansted, in twenty-two minutes. This method of transportation resulted after a survey showed that 64 percent of British air travelers originated their travel from within five miles of Victoria Station.

The English system may be easier to solve than the problem in America, where two thirds of all air passengers come from the suburbs. One idea to solve this muddle is the Skylounge, which would drive around town picking up passengers and, when filled or ready for airport departure, be airlifted like a helicopter to the air terminal. This idea is being studied in Los Angeles.

Another idea is the monorail. Still another, the underground rapid transit system for airport purposes only.

And yet all the ideas fall flat on their faces because of one awful moment when fog descends over the airports and all flying and all access traveling ceases, with utter confusion taking the reins. So far, little if anything is being done to cope with fog which in the new decade of the jets will continue to paralyze scheduled flying.

Attempts to bring smoother service and safer operations into the airport area by designating a certain number of flights during high peak traveling periods of morning and late afternoon do not solve the muddle, for the success of such operations is dependent entirely on good weather and good visibility. When weather minimums start to fall, when air pollution and fog drift over the complexes, all the spaced-out flight planning collapses.

Entirely new concepts in size and structure of airports, the location, the access surface system, and complete automation may someday solve the problem. Until that time arrives, the increase in traffic volume will send the accident rate soaring to new levels.

Will anyone pay attention to the warning? As far back as 1951, General James Doolittle warned communities not to build apartments or motels near airport approaches, in an attempt to help solve the safety problem of that time. Then, sixteen years later occurred the very thing he was fearful of.

On March 30, 1967, a Delta Air Lines DC-8 with six crew members aboard for regulation check-ups took off from New Orleans International Airport for a routine operation that required a circling flight to approach and land on the same runway. With Delta's top check-pilots on board and an FAA inspector to supervise the procedures, the DC-8 faltered.

On its final approach to the runway, over homes and other buildings, the jet began to sideslip to the left. The crew's alarm was shown later in the cockpit recording instrument and eight seconds later the aircraft struck a forty-foot-high tree, then two more trees which split the wings and caused fuel to stream from the aircraft over the houses and street. The DC-8 next slashed through the corner of a house, struck a panel truck on a street, struck the ground by another house, broke up, totally destroyed another house, sprayed flaming kerosene over other houses in the vicinity and continued onward. It ran along the ground, smashing everything in its pathway, traveled over a railway embankment and finally came to rest against the Hilton Inn Motel.

Here explosions ripped the aircraft into thousands of flaming missiles and surrounding buildings caught fire. The crew perished. Thirteen persons died in their homes and in the motel.

The case was an error in human judgment on the flight deck. But what about the error in judgment in permitting homes and a motel and other occupied buildings to be built so close to an airport's runways. No one mentioned this aspect of the case. It is recalled here as a lesson to airport planning of the future.

13

Jumbojets and Supersonics

So little time is left.

The problems of the first decade of the jets have not been solved—neither air traffic control, nor passenger safety, nor weather, nor instrumentation, nor training, nor airports.

The problems should have been solved before the introduction of the behemoths of the second jet decade which will be dominated by the jumbojets, airbuses and supersonics. Only a crash program by the Federal government, operators, manufacturers of aircraft, town planners and road commissions can bring a semblance of order and safety out of threatened chaos.

Already jets have been stretched longer to hold up to 200 passengers. Jumbojets with passenger-carrying potentials of 500 people will be flying in 1970 and a few years later, if the manufacturers and airlines get their way, jets carrying up to 1,000 passengers will be flitting from city to city creating more noise, more confusion, more accidents and greater amounts of pollution.

It is not true that jumbojets and supersonics will solve air traffic problems. They will only be able to satisfy the increase in use of air transportation and the need to get from one place to another in a shorter period of time. At the present state of air traffic and airport congestion they won't even be able to satisfy these requirements, much less safety, noise and other factors, unless something is done quickly to alleviate the present muddle.

There are many people at this stage in the game, who

wonder whether it will be at all safe to fly in such tremendous airliners. The answer is yes. Jumbojets and supersonics will have the aerodynamic and structural ability to handle the loads assigned to them and will be powered by fantastic new turbine engines that will scoot them off the airports and across the skies as rapidly as the conventional jets. They will be much noisier because of the larger engines, twice as powerful as today's. Supersonics will spread the sonic boom across the flight path wherever they travel and there is some thought of limiting their flights to transocean only. But this remains to be seen. The United States is not going to watch other nations build and fly supersonics where they please without attempting to do the same. We may as well face it: supersonics will be permitted to fly across this country as soon as the Boeing SST gets airborne sometime toward the end of the next decade.

There is an added safety problem, however, when considering the jumbojets. The relationship between weight, size of the wingspan and turbulent air means that jumbojets may be forced to fly only in air corridors that are clear of severe weather. This should not be too difficult to establish—*after* the crash of one Moby Dick with 500 to 1,000 lives snuffed out at one instant. Let us hope that it doesn't happen in the State of Maryland where the law says that an autopsy must be performed on every person killed in a plane crash. There would not be enough pathologists in the country to conduct a single crash investigation in the state.

The "big daddy" of the new crop of jets is the C-5A, dubbed the Galaxy, and built by Lockheed Aircraft Corporation at Marietta, Georgia. Lockheed, which concentrated its efforts on the Electra turboprop when the jet age was dawning and lost out in the race to Boeing and Douglas, was again severely slapped by Boeing in the supersonic design race. But the C-5A was expected to bail Lockheed out of the commercial doldrums and get the company back into passenger service, where it had had such a long and creditable career with the Lodestars and the famous Constellation series.

The C-5A was intended for the military. Its successor in

passenger service will be a modified version. Built for the Air Force which ordered fifty-eight of the giants, the C-5A is designed to carry a staggering load of 100 tons over 3,100 miles or fifty tons over 6,325 miles. It would have taken only twelve such jumbos to conduct the entire Berlin airlift of 1948–49, an operation that required round-the-clock operations of 224 aircraft of that period.

The C-5A has been undergoing flight tests since June 1968 and its performance in the air has been impressive. It will be shuttled into service in mid-1969, with further deliveries scheduled well into the 70s. It is almost the length of a football field, 246 feet long, with a wingspan of 223 feet. It rises six stories high and squats on twenty-eight low-pressure tires that permit it to land on a field muddy enough to stall the family car. It is the first aircraft to use computerized systems to detect in-flight malfunctions.

The inside is a vast hall which can be used for cargo or a passenger-cargo mixture. This cabin is thirteen feet high and could accommodate an eight-lane bowling alley on its main deck. An upper deck holds an additional seventy-five persons as well as two spare flight crews. This $20,000,000 machine can remain in the air indefinitely with refueling, and it is large and structurally tough enough to launch missiles. As a passenger model it can haul 800 people with all their baggage from coast to coast or across the oceans at a speed close to two thirds the speed of sound. As one observer remarked as he watched its first flight, "It reaches the bounds of credibility." A Lockheed designer standing nearby answered: "There is no such thing as a maximum plane . . . they will get bigger and bigger . . . and why not?"

The gross weight of the C-5A is 728,000 pounds. By contrast the 707 intercontinental jet, generally considered a monster today, has a gross takeoff weight of 328,000 pounds. But useful payload is the key statistic of today's commercial transport business, and the C-5A (which will be called the Lockheed 500 as a passenger plane) will be able to carry 141 tons, three times more than the biggest air freighters of today.

In the meantime, Lockheed is planning a slightly smaller version to fill the interim between the present jets and the 500. This newest jumbo will be known as the Lockheed L-1011. Since it takes seven years from drawing board to commercial service, the Ten-Eleven will not be in the air until 1971.

Referred to as the Lockheed airbus, designed principally to tap the riches of hub commuter travel where 70 percent of all air revenue is amassed, it has a range of 3,160 miles, will carry a payload of 56,200 pounds—including 256 passengers and 5,000 pounds of baggage—and will cruise between 31,000 and 35,000 feet at 565 miles an hour. Fuel required for a maximum flight will be 100,000 pounds of kerosene with an additional 25,000 pounds in the reserve tanks.

The maximum gross takeoff weight of 409,000 pounds will permit the aircraft to use an 8,750-foot runway with a thirty-five-foot-high obstacle at the end. It will carry a full load of passengers from New York to Los Angeles or from San Francisco to Honolulu. For routes of much shorter duration, the aircraft will be able to operate off 7,000-foot runways such as La Guardia and Chicago's Midway airports.

The Ten-Eleven will need 5,650 feet of concrete to land and its maximum set-down weight is figured to be 348,000 pounds. This means that, if an emergency takes place after takeoff, this aircraft will have to dump some thirty tons of fuel before it can safely set down again. It will be powered by three Rolls-Royce by-pass turbofan engines, rated at 40,000 pounds of thrust each or better. Two engines will be mounted on pylons under the wing while the third will be anchored into the rear cone of the fuselage, not unlike the present 727.

Lockheed's sales pitch is that the Ten-Eleven will have significantly lower noise levels than current transports because of the reduction in fan whine, a quieter concept in nacelle (engine covering) design, and a power cutback that will not endanger the stability of the aircraft. Its wingspan of 155 feet four inches will not require wider runways than are currently in use. There is no secret that Lockheed intends to use the experience of the C-5A to incorporate it into the Ten-Eleven.

Douglas Aircraft Division is entering the jumbojet era in a series of giant leaps which started with the stretching of the DC-8s to increase passenger capacity from about 120 per jet to 243 passengers. The next step is the design and introduction of the DC-10 which will carry up to 600 passengers and be ready for certification about 1973. Douglas is also in the design stage of a 1,000-passenger jet which has already placed General Electric and Rolls-Royce in a nose-to-nose battle to come up with even more powerful engines than are now in the design stage.

Boeing, Number One in passenger aircraft manufacturing, announced in early 1966 that the sale of its first jumbojet, the Boeing 747, had been made to Pan American World Airways. The first 747 will be delivered about September 1969 and thereafter Boeing will deliver close to nine completed aircraft every month. Pan Am hopes to make its first transoceanic flight in December 1969 and has already sold the 366 plush-class tickets.

Publicity releases for the first flight offer a faster and higher flight than is achieved by present jets. There will be more legroom, elbowroom, headroom and hiproom, with bigger cabins, observation areas with "pilot-eye view," private compartments for business conferences, and staterooms with beds —everything for those in a hurry.

Despite the glowing plans, it is hard to see any air line using the 747 as anything less than what it is intended to be, a huge skybus that can transport 400 people (and more than that) long distances at a lower cost—though cost is a touchy subject with wages bound to increase and airport charges and maintenance costs moving upward rapidly. The FAA has already planned large increases in jet fuel taxes to help pay for the cost of airport improvements. When Boeing and Douglas began to talk about low fares with their skybuses, and advertised lower fare estimates in their promotion, the airlines objected. A spokesman for the industry said: "It's all right for Douglas and Boeing to talk operating costs and seat mile costs but the carriers will determine what tariffs they will seek."

As a size comparison of the jets, it may be interesting to note that the present Boeing 707 is 153 feet long, the Douglas DC-8 stretch-jet is 187 feet, the Boeing 747 will be 229 feet, and the coming supersonics will be close to 300 feet.

How big is big? The main landing gear of the 747, as an example of size, will be twenty feet long, thirty-seven inches in depth, will weigh 1,760 pounds. This introduces another problem. Are today's jet runways strong enough to withstand such shock and weight? A survey is now being conducted to find out, and if the runways are not thick enough and strong enough, another safety problem raises its head in an already complex situation.

The safety worry concerning the jumbojets is their great size and passenger-carrying capacity. With the supersonics, however, the safety problem is much more complex and some experts have claimed that supersonic flying will be downright dangerous. The supersonic speed alone makes the flight crews unaware of impending collisions with weather or other aircraft. Hail and precipitation along the flight path will have to be avoided and these nuisances have been located by the U.S. Air Force at heights over 100,000 feet. Clear Air Turbulence has been located with nagging regularity over 50,000 feet where the supersonics will fly. Turbulence could tear a supersonic apart.

Passenger and crew safety may also be endangered by solar radiation from sunspots and from galactic radiation. Meteor collisions are also possible at such heights and these could cause punctures of the fuselage and depressurization of the cabins. Supersonic pilots will have to wear oxygen masks. The passengers will have about thirteen seconds, or less, to get their masks in place, or die.

The first supersonic to fly will probably be the British-French Concorde. We say probably because the Russians are slightly ahead of the Concorde and in all likelihood will not go through a lengthy time of flight testing and therefore could be in service from Moscow to New York and to Montreal by 1971. The Concorde, which is in its primary testing phases in

England and France, will not likely be ready for passenger service until 1972. The Boeing supersonic which has met with serious design problems and has been changed from a variable wing to a delta wing, like all the others, may not get off the ground until 1977. If its costs keep mounting it may never fly.

The Concorde is a luxury aircraft and only the well-heeled will be able to afford its fares. It will carry 136 passengers and will fly from London to New York into the prevailing westerly winds in three hours and seventeen minutes as compared with today's 707 time of seven hours and thirty-five minutes. The Concorde will fly from Paris to Buenos Aires in slightly over seven hours and from London to Sydney in eleven hours and twenty minutes as compared to twenty-three hours in a standard jet. It will be possible to attend the races at Fécamp and then jet to New York—all in the same day.

Passengers will not be aware they have passed through the sound barrier unless the pilot tells them so, such is the sophistication of the delta wing design. But if the passengers are unaware they have broken the sound barrier, many on the ground below will certainly know it.

The first sight of the Concorde is thrilling. At Filton, the British Aircraft Corporation has a surprise manner in which their pride is demonstrated. You are taken into a tremendous hangar which is in complete darkness. From somewhere a small spotlight comes on and this is followed by several other dim lights and then the first shape of this monstrous plane, 191 feet long, 38 feet high, standing on its stiltlike undercarriage, looms out of the darkness above. Next the interior lights glow from the line of windows and the needle nose stands out clear ahead of the thin streamlined body. Next the wings become gradually illuminated, tremendous wings, with a span of 83 feet, 10 inches, that seem to fall downward and away from the aircraft. Then the rear assembly and the long pointed tail of the fuselage are illuminated and the Concorde takes shape.

One's first thought is, how will an airplane this size ever fly? How will it take off in absolute safety? How will it land at such high speeds?

The Concorde builders answer that the Concorde will be no faster nor slower than a conventional jet during landings and takeoffs. Asked what they thought about the safety aspects involving solar flares, sonic booms, turbulence, and so on, they replied: "Muddled thinking—Concorde is a safe aircraft."

14

Safety and Acceptability

Air safety is everybody's business.

Yes, even the man who has never flown and perhaps never will, should be concerned with the problems of air safety. One of his family may die some day in a midair collision or his house may be obliterated by a jet trying to land while observing noise abatement procedures and flying into an inadequate airport whose expansion he fought and whose relocation he was not remotely concerned with.

The future of aviation depends upon acceptability by the public who do not fly and by the passengers who do. A tremendous educational program must be started at once to acquaint everyone with the problems of air transportation because its future, now endangered by apathy on one side and bitterness on the other, is linked to community, industrial and national growth for all time to come.

To make safety work, air lines, international air transport groups and governments must be made to realize that "acceptable fatal accidents" are not acceptable. They must face facts and stop hiding behind statistics that do not tell the truth.

For instance, ICAO and the operators show that in 1961, 256 persons died in jet accidents on scheduled airlines. This is not so, 310 persons died. The crew members were deliberately left out, in order to make the death toll look lower. In 1962, ICAO shows that 424 persons were killed, when in reality 610 persons died. In 1963, ICAO and the airlines show by their figures that 347 were killed when in fact 383 died in crashes.

As a further example of this type of reporting, from 1961 until 1965, ICAO and its members and government agencies caught in the statistical web of misinformation, show that 1,414 persons died in jet crashes. They were passengers, it is true. But others died in the same crashes bringing the total to 1,732 dead. This makes the death toll in jet disasters more than 22 percent higher than is officially released.

This use of only revenue passenger deaths is a ploy to keep the death statistics within the "acceptable" fatalities limits which the airlines like to use. It is a distasteful way to create images that just do not exist. The FAA and all other government enforcement agencies should immediately scrap the "passenger mile criteria" and permit safety standards to be assessed by facts rather than by propaganda.

Federal authority must allocate the billions of dollars necessary to rebuild and improve the nation's commercial airports. To expect cities that are sapped by strife and turmoil to contribute to this expensive program is fallacy. To expect states and counties to share the costs of such expensive projects is wishful thinking. Most air travelers come from the suburbs that are today scattered over vast areas. Their living and traveling habits may be the clue to the location of future airports but national transportation is still a Federal problem, including the construction of access superhighways and high-speed rapid-transit systems to feed such complexes.

The airport program must be completed within five years before utter chaos takes over. But first things first. The most important airports should be those of the hub areas: New York City, Chicago, Los Angeles, Atlanta, Miami, Toronto, Detroit, and so on down the line in volume of traffic. Each airport should be built a good fifty or sixty miles from city limits and be connected by two forms of transportation, a 200 mile-an-hour rapid-transit system and a superhighway system designed and routed specifically for the airport.

The complex itself should have the individual air terminals

of the carriers dotted around the perimeter and connected to each other and to the parking lots and to the rapid-transit system by a snappy shuttle train which would serve everything on a five-minute basis, longer during the slower transportation periods, if such a thing exists in the future.

The runways need not be too long, about 12,000 feet would be sufficient for jumbojets and supersonics but the overrun and undershoot areas should be at least a half a mile in length and paved. Between the runways—which will be numerous depending on traffic conditions—man-made lakes of not more than three feet in depth must be constructed so as to act as cushions for missed-landings and aborted takeoffs, and to act as fire depressants. These bodies of water should be heated during bad weather conditions to dissipate fog. The warm water could also be used as a beacon to infra-red sensing devices in each airliner, a system which would actually outline the runways between the water better than the outlines of radar.

Car parking areas could be underground and should be readily accessible to the shuttle train system, not more than three hundred feet to any one platform.

Baggage, always a headache and an irritating problem to airlines and passengers alike, should be carried separately from the jet passenger aircraft and should be placed aboard jet cargo airliners to fly ahead of the jumbo passenger planes and have the baggage waiting on the passengers' arrival. One cargo jet, for instance, could pick up the baggage in New York of several carriers leaving for Los Angeles and, augmented with fast-cargo shipments, could easily be in Los Angeles well ahead of the passenger flight. This procedure would help eliminate the hazard of sabotage in passenger aircraft.

The airport complex of runways and buildings and parking areas should be completely surrounded by a forest of trees, one mile in depth to absorb noise and pollution and to act as a cushion for possible approach crashes. The forest concept could be designed as an attractive green belt for the district and be laced with picnic areas and observation grounds where visitors could watch the thrilling sight of supersonics and other giant airliners landing and taking off.

Safety must be the number one item of the flight deck. Upgraded quality is needed for all instrumentation. There are, at present, just too many instances of faulty altimeters and other instruments, some of them badly installed and difficult to read.

To protect this tremendous commercial operation, the Federal authority must construct new airports and upgrade others for private and business aircraft, and air traffic corridors should be allotted to each group.

All flights, commercial, business and private cross-country flying should be made under precise radar surveillance at all times of the day and night in all the skies over the United States and possibly Mexico, Canada and the adjacent islands of the Caribbean. For safety's sake all training programs should be conducted from airports specifically marked for training. Aircraft used must be equipped with radio and up-to-date navigation systems. This would lower the risk of mid-air collisions.

Weather reporting must be updated for the supersonic age. At present it is often slipshod because of the lack of funds to carry on an intensive survey of the atmosphere at all flight levels. Every airport should be radar equipped. This could be of additional assistance in compiling local and national weather forecasts.

Airliners should not be permitted to fly into or from airports that do not meet the minimum standards set down by the FAA and the Air Line Pilots Association as to runway lengths, fire protection, safety rescue systems and overrun areas.

Maintenance must be sharply upgraded by the air lines. Radar should be improved. Aircraft fuel should be contained in spill-proof indestructible tanks. Passenger compartments should have an exit for every twenty persons with clear aisles unencumbered by seats and carry-on luggage. Every aircraft should have a sprinkler system of nontoxic fire suppressant material. Explosive exits should be embedded into the fuselage to explode outwardly to free trapped passengers as well as crew members.

No aircraft should be certified for public use unless it can be evacuated in 45 seconds.

A qualified member of the crew should be near each emergency exit. High-powered lights and flashing strobe lamps should show the way if smoke or other fumes cause a loss of visibility. Seating should be arranged so that old people and mothers with youngsters can be helped quickly and efficiently to safety in aisles uncluttered by seats and carry-on baggage.

Flying is a delightful way to travel. It should be safe and can be safe. It needs strength and courage to make it safe.

If you agree that it can be done, join the crusade. In the meantime, fasten your seat belts, cross your fingers as the pilots do in the hope of avoiding a midair collision and have a happy landing.

Index

A

B

D

E

F

G

H

L

Mc

M

N

O

P

R

S

U

V

W

Z